ROYAL NAVY STRATEGY IN THE FAR EAST 1919–1939

CASS SERIES: NAVAL POLICY AND HISTORY
ISSN 1366–9478

General Editor: Geoffrey Till

This series consists primarily of original manuscripts by research scholars in the general area of naval policy and history, without national or chronological limitations. It will from time to time also include collections of important articles as well as reprints of classic works.

1. *Austro-Hungarian Naval Policy, 1904–1914*
   Milan N. Vego
2. *Far-Flung Lines: Studies in Imperial Defence in Honour of Donald Mackenzie Schurman*
   Edited by Keith Neilson and Greg Kennedy
3. *Maritime Strategy and Continental Wars*
   Rear Admiral Raja Menon
4. *The Royal Navy and German Naval Disarmament 1942–1947*
   Chris Madsen
5. *Naval Strategy and Operations in Narrow Seas*
   Milan N. Vego
6. *The Pen and Ink Sailor: Charles Middleton and the King's Navy, 1778–1813*
   John E. Talbott
7. *The Italian Navy and Fascist Expansionism, 1935–1940*
   Robert Mallett
8. *The Merchant Marine in International Affairs, 1850–1950*
   Edited by Greg Kennedy
9. *Naval Strategy in Northeast Asia: Geo-strategic Goals, Policies and Prospects*
   Duk-Ki Kim
10. *Naval Policy and Strategy in the Mediterranean: Past, Present and Future*
    Edited by John B. Hattendorf
11. *Stalin's Ocean-going Fleet: Soviet Naval Strategy and Shipbuilding Programmes, 1935–1953*
    Jürgen Rohwer and Mikhail S. Monakov
12. *Imperial Defence, 1868–1887*
    Donald Mackenzie Schurman; edited by John Beeler
13. *Technology and Naval Combat in the Twentieth Century and Beyond*
    Edited by Phillips Payson O'Brien
14. *The Royal Navy and Nuclear Weapons*
    Richard Moore
15. *The Royal Navy and the Capital Ship in the Interwar Period: An Operational Perspective*
    Joseph Moretz
16. *Chinese Grand Strategy and Maritime Power*
    Thomas M. Kane
17. *Britain's Anti-Submarine Capability, 1919–1939*
    George Franklin
18. *Britain, France and the Naval Arms Trade in the Baltic, 1919–1939: Grand Strategy and Failure*
    Donald Stoker
19. *Naval Mutinies of the Twentieth Century: An International Perspective*
    Edited by Christopher Bell and Bruce Elleman
20. *The Road to Oran: Anglo-French Naval Relations, September 1939–July 1940*
    David Brown
21. *The Secret War against Sweden: US and British Submarine Deception in the 1980s*
    Ola Tunander
22. *Royal Navy Strategy in the Far East, 1919–1939: Planning for War against Japan*
    Andrew Field

ROYAL NAVY STRATEGY
IN THE
FAR EAST
1919–1939

Preparing for War against Japan

Andrew Field

FRANK CASS
LONDON • NEW YORK

First published in 2004 in Great Britain by
FRANK CASS
2 Park Square, Milton Park,
Abingdon, Oxon, OX14 4RN

Simultaneously published in the USA and Canada by
FRANK CASS
270 Madison Ave, New York NY 10016

Frank Cass is an imprint of the Taylor & Francis Group

Transferred to Digital Printing 2006

Copyright © 2004 A. Field

British Library Cataloguing in Publication Data

Field, Andrew
 Royal Navy strategy in the Far East 1919–1939: preparing for war against Japan. – (Cass series. Naval policy and history; 22)
 1. Great Britain. Royal Navy – History – 20th century
 2. Sea-power – Great Britain – History – 20th century
 3. Navy-yards and naval stations, British – Asia – History – 20th century 4. Great Britain – History, Naval – 20th century 5. Great Britain – Military policy – History – 20th century 6. Great Britain – Defenses – History – 20th century 7. Great Britain – Foreign relations – Japan 8. Japan – Foreign relations – Great Britain 9. Great Britain – Foreign relations – 1910–1936 10. Great Britain – Foreign relations – 1936–1945
 I. Title
 359′.03′0941′09042

ISBN10: 0-7146-5321-7 (hbk)
ISBN10: 0-415-40775-3 (pbk)

ISBN13: 978-0-7146-5321-1 (hbk)
ISBN13: 978-0-415-40775-5 (pbk)

ISSN 1366-9478

Library of Congress Cataloging-in-Publication Data

Field, Andrew
 Royal Navy strategy in the Far East 1919–1939: preparing for war against Japan / Andrew Field.
 p. cm. – (Cass series – Naval policy and history; 22)
 Includes bibliographical references (p.) and index.
 ISBN 0-7146-5321-7 (cloth)
   1. Great Britain. Royal Navy – History – 20th century. 2. Great Britain – History, Naval – 20th century. 3. Great Britain – Foreign relations – Japan. 4. Japan – Foreign relations – Great Britain. 5. East Asia – Foreign relations – Great Britain. 6. Great Britain – Foreign relations – East Asia. 7. Great Britain – Foreign relations – 1910–1936. 8. Great Britain – Foreign relations – 1936–1945. I. Title. II. Series.
 VA454.F485 2003
 359′.03′094109041 – dc22

All rights reserved. No part of this publication may be reproduced, stored in or introduced into a retrieval system or transmitted in any form or by any means, electronic, mechanical, photocopying, recording or otherwise, without the prior written permission of the publisher of this book.

Typeset by Tradespools, Frome, Somerset

Printed and bound by CPI Antony Rowe, Eastbourne

Contents

List of Illustrations	vi
List of Maps	vii
List of Tables	viii
Series Editor's Preface	ix
Preface	xi
Acknowledgements	xiii
Introduction	1
1 The Influence of a Far Eastern Strategy on British Naval Policy	19
2 A Far Eastern Strategy: War Memorandum (Eastern)	48
3 Admiral Richmond and War Memorandum (Eastern)	74
4 Developing the Far Eastern Strategy: War Memorandum (Eastern) and Changing Circumstances, 1931–41	97
5 Battle Fleet Tactics and a War in the Far East	123
6 The Royal Navy's Strategic and Tactical Exercises	158
7 Japanese Naval Strategy and Tactics in the Far East	183
8 Main Fleet to Singapore: The Sinking of HMS *Prince of Wales* and HMS *Repulse* and the End of War Memorandum (Eastern)	213
Conclusion: War Memorandum (Eastern) and the Royal Navy's Strategic, Operational and Tactical Development	230
Select Bibliography	250
Index	267

Illustrations

		page
1.	The traditional view of British sea power; British battleships in line ahead.	33
2.	A Sopwith *Cuckoo* drops a torpedo on the Atlantic Fleet in Portland Harbour, 6 September 1919.	34
3.	HMS *Queen Elizabeth*.	34
4.	HMS *Resolution*.	34
5, 6.	Two views of HMS *Courageous*.	35
7.	One of the Royal Navy's *County*-class cruisers.	36
8.	HMS *Dauntless*.	36
9.	HMS *Leander*.	36
10.	HMS *Southampton*.	37
11.	HMS *Express*.	37
12.	*Tribal*-class destroyer.	37
13.	Retired Admiral, Percy Scott's 1920s prophetic view of a naval battle of the future.	125
14.	Japanese battleship *Mutsu*.	198
15.	The battleship *Fuso*.	199
16.	The fast battleship *Kongo*.	199
17.	The aircraft carrier *Kaga*.	200
18.	The light cruiser *Yura*.	200
19.	Improved versions of Japan's first Treaty cruisers, the *Furutaka* class.	201
20.	Japanese cruiser *Ashigara*.	201
21.	Inter-war Japanese destroyer *Yusuki*.	202
22.	*Shikinami* of the *Fubuki* class.	202

Credits: photographs 2, 5, 6, by courtesy of the Fleet Air Arm Museum; 13, from P. Padfield, *Guns at Sea* (Hugh Evelyn, London, 1973), p. 283; 14, 15, by courtesy of A.J. Cashmore; 16 20, 21, by courtesy of Maritime Photo Library; 17, 18, 19, 22, by courtesy of the National Maritime Museum; remaining photographs are from the author's collection.

Maps

		page
Map 1:	The Major Battle Fleet Strength of the World, 1931	20
Map 2:	War Memorandum (Eastern), The Route to the East	52
Map 3:	War Memorandum (Eastern), 1924–31	65
Map 4:	The Naval Balance in 1941	223

Tables

		page
Table 1:	Oil fuel requirements and proposed supply arrangements in the east, 1921	55
Table 2:	Oil storage proposals, 1923	61
Table 3:	British battle fleet in cruising formation, 1924	136
Table 4:	Rates of hitting table	145
Table 5:	Number of necessary Non-Vital Hits by 12-inch gun and above to sink a battleship	146
Table 6:	Effects of 15-inch gunnery on cruisers and unarmoured vessels	146
Table 7:	Effects of 6-inch gunnery on cruisers and unarmoured vessels	146
Table 8:	Admiralty War Game Hits per Gun per Minute table, 1929	147
Table 9:	Japanese naval strength	195

Series Editor's Preface

The long story of the shifting interwar policy of the Royal Navy towards the Far East ended with the sinking by the Japanese of the *Prince of Wales* and the *Repulse* on 10 December 1941. For the British this was a tragic event which signalled the end of an era of strategic dominance over the area; when taken alongside the successful Japanese assault on the US Pacific fleet at Pearl Harbour, that defeat also appeared to signal the end of the battleship era in naval warfare. Put together, these two events of December 1941 seem to have amounted to a transformation, a revolution in military affairs. Everything afterwards would be and was different from what had gone before.

This book explores and analyses the background to these events and shows, however, that things were not so simple, and the Royal Navy was perfectly aware of the real challenges it faced and accordingly evolved a pragmatic and adaptive strategy to deal with them. December 1941 was not a bolt from a clear blue sky, destroying the blissful ignorance of a naval generation too blinkered to be aware of what was happening around them.

Instead, those events need to be seen as the end of a long and for the most part perfectly conscious process of adaptation to a steadily deteriorating situation. At the level of grand strategy, the notion of building up Singapore, and sending out a main fleet from European waters, when circumstances permitted, seemed the only way in which the Admiralty could balance its resources and its commitments. During the 1920s, the Royal Navy was confident that this would secure British interests in the Far East.

But during the 1930s with the rise of serious German and Italian threats in European waters, such aspirations had to be reduced. The main fleet became smaller, the time interval before Singapore could be relieved longer. The critical assumptions that the Royal Navy made (that the US Fleet would continue to occupy the attention of the Japanese, that the Japanese would be launching their assault on Singapore from the North West Pacific, rather than French Indo-China, that significant forces including a carrier would be available) gradually became unstitched. In the end, what materialised was a flying squadron that was essentially intended to act as a deterrent. But by the time it arrived, the actions it was intended to deter were already taking place and, by a mixture of possibly excessive offensiveness (in striking contrast to the supine indecision so frequently said to characterise the British defence of Malaya) and sheer bad luck, its bluff was fatally called.

But as Andrew Field so convincingly shows, all aspects of this issue in fact were meticulously and thoughtfully discussed for the twenty years that went before. British responses were sensible and considered – if ultimately a hopeless bid to square the circle.

But these discussions also provide a crucial window into British naval thinking about the British approach to naval warfare. In their discussions about what the main fleet, or the flying squadron, would actually do at every level of war when it finally arrived on station, the Royal Navy demonstrated that it was not the conservative, hidebound battleship dominated force as it used once to be portrayed. This was a very different navy from the one that had fought Jutland. The fundamental problem was that it simply did not have the assets to fight and win a major naval war in the Far East and in Europe at the same time. And this critically undermined its ultimate capacity to defend British interests in the Far East. In his discussion of these melancholy developments, Andrew Field does ample justice to the importance of his topic.

Geoffrey Till

Preface

Between 1919 and 1931, in the absence of any real European threat, the Royal Navy's strategical and tactical planning was focused on Japan, only shifting towards first Italy and then Germany from 1933, as the political situation in Europe deteriorated. The purpose of this book is to consider the background to this, the challenges to the Royal Navy's traditional naval superiority posed by Japan and the responses to them and the shift in naval thinking to the Pacific Ocean as the future area of conflict, both factors culminating in the building of the Singapore naval base and the development of a Far Eastern strategy, War Memorandum (Eastern) of 1924, the strategy to fight the Japanese.

Long-seated mistrust of Japan was used as the justification for a major naval base at Singapore, and the development of War Memorandum (Eastern) in 1924. As it was impossible to have a fleet permanently based at Singapore, the strategy had to encompass the logistics of sending and maintaining a fleet to the Far East, as well as how best to use it to destroy the Japanese Fleet. How the battle fleet would fight was also considered. Battle Instructions were developed which encouraged the use of initiative, with the aim of defeating the Japanese.

The changes in Britain's strategic priorities will also be discussed – changes which culminated in the decision to send the *Prince of Wales* and *Repulse* to Singapore, and their ultimate sinking by Japanese aircraft in December 1941 – and the exercises that were used to test out doctrine and tactics.

As a strategy, War Memorandum (Eastern) had many flaws but its real importance lay, perhaps, in the fact that it was the common language between the strategists and the tacticians and the yardstick by which the Royal Navy was able to measure its efficiency for any future war.

Acknowledgements

This book started as an MPhil. thesis and so I would like to start off by acknowledging all of the help and support given to me by my tutor, Professor John Gooch, when I was working on this. I must also acknowledge the support of Jock Gardner of the Naval Historical Branch, London, and Professor Eric Grove of the University of Hull, both of whom have encouraged me in my research and given freely of their time and expertise, advising and helping me shape my ideas over the past ten years.

I must also thank many others who have helped me along the way. These include Christopher Bayne for allowing me to make use of the papers of his father, Captain R.C. Bayne, RN, Captain Christopher Page, RN, of the Naval Historical Branch, for permission to use documents from their archive, Roderick Suddaby of the Imperial War Museum, London, and Commander Mike Mason, RN, Head of Defence Studies (Royal Navy), for their willingness to grant permission to use materials in their keeping.

I must also thank Mr John Wenzel, the Director of HMS *Belfast* and his staff for explaining the principles behind the Admiralty Fire Control Table to me, D.M. Stevens, the Director of Naval Historical Studies, Canberra, for materials relating to the Australian Navy and the late Brian Ranft for his letter about Admiral Beatty. In addition, I must also thank Gareth Simon, for his permission to use materials from his extensive *Pallas Armata* reprints.

I am also indebted to the Master, Fellows and Scholars of Churchill College in the University of Cambridge, especially the staff of the Churchill Archive, for all of their time and trouble, and for permission to use and quote from the papers of Admiral Sir Frederick Charles Dreyer, Captain The Honourable R.A.R. Plunkett-Ernle-Erle-Drax and Captain Stephen Roskill.

I also wish to thank the National Maritime Museum, Greenwich, for allowing me to use and quote from the papers of Admiral Sir Herbert Richmond, and for helping me with the photographs of Japanese warships, and Matthew Sheldon, the Curator of Manuscripts at the Royal Naval Museum, Portsmouth, for introducing me to the 1925 Lecture Précis notes and for permission to use these.

Professor Christopher Bell of the United States Naval War College, who has also worked on the same topic, very generously sent me a copy of *The Royal Navy, War Planning and Intelligence between the Wars*, and offered some useful comments and ideas. Evelyn Cherpak, also of the United States Naval War College, granted me permission to use documents from their archives. I wish to thank Clark Reynolds of the University of South Carolina, for a copy of Minoru Genda's unpublished monograph, 'Evolution of Aircraft Carrier Tactics of the Imperial Japanese Navy'.

I would also like to thank the Maritime Picture Library, the Fleet Air Arm Museum and A.J. Cashmore for the illustrations, and Ian Bolton of the Anatomy Visual Media Group for his superb work on the maps. I have tried to obtain permission for all the copyright material I have used but, if I have failed, it will be rectified in future editions, on application by copyright-holders to the publisher or to me.

Finally, I must acknowledge the continued support given to me by my Head Teacher and colleagues, who have had to cover for me when I have disappeared to various conferences and lectures. Thanks are also due to Andrew Humphrys of Frank Cass for his guidance and support, Sheila Kane for her copy-editing, and, most importantly, my wife Pauline, for her advice, help and encouragement along the way. She is the one who has had to listen to all of my moans, and ideas, right from the very start.

As any researcher knows, this book has been a collaborative effort. I could not have completed the book without all of the above help and support, or without the permission of other publishers to allow me to quote from their authors' works. However, any omissions or inaccuracies that occur within this book are the sole responsibility of the author.

Introduction

On 10 December 1941, three days after the attack on Pearl Harbor disabled the US Pacific Fleet, Force Z, the Royal Navy's battleship *Prince of Wales* and the battlecruiser *Repulse* were sunk by Japanese aircraft. On 25 December the colony of Hong Kong fell to Japanese forces and on 15 February 1942, Singapore, with its extensive naval facilities, surrendered. Twelve days later, at the Battle of the Java Sea, Allied naval forces were annihilated by Japanese forces. By April 1942, Japanese aircraft carriers under Admiral Nagumo, the commander of the Pearl Harbor raid, were at large in the Indian Ocean, sinking the small aircraft carrier *Hermes* and the cruisers *Cornwall* and *Dorsetshire*, whilst the British East Indies Fleet under Admiral Somerville had to manoeuvre to avoid action on unfavourable terms. Darwin, in the Northern Territory of Australia, had already been bombed, on 19 February, and in May 1942 Japanese midget submarines attacked shipping in Sydney Harbour, albeit with limited success. British Far Eastern naval strategy, carefully crafted and developed in the years following the First World War, was shattered by these disasters. British forces were on the defensive everywhere in the region and it would be 1945 before a Royal Navy fleet was again active in the Pacific. By then the United States Navy reigned supreme, the largest and most powerful naval force in the world, and the end of Britain's imperial presence in the region was inevitable.

To understand why Britain's Far Eastern naval policy appeared to be so unsuccessful it is necessary to go back to the end of the Great War, and examine the roots of the strategy. Between 1919 and 1930 the Royal Navy became increasingly concerned with imperial security in the Far East and once the Washington Naval Treaty had been signed in 1921, and naval parity with the United States Navy achieved, Japan became the focus of British strategic planning, and War Memorandum (Eastern) was developed to meet the challenge in the East.

With its emphasis on a decisive fleet action, War Memorandum (Eastern), was a perfect reflection of the characteristics of the First Sea Lord, Sir David Beatty and his supporters (or, as Andrew Gordon prefers to refer to them in *The Rules of the Game*, 'ratcatchers'), an aggressive interpretation of naval warfare and naval strategy.[1]

The roots of this emphasis lay in the recently won war against Germany. The Royal Navy had entered the Great War with the largest battle fleet, but one that was constrained by a rigid doctrine of command which discouraged the use of initiative and judgement. In Gordon's words:

> To reduce Jutland to a sound-byte, in spite of losing two of his battlecruisers through partly self-inflicted causes, Beatty succeeded in drawing the German Fleet into the killing zone in front of the Grand Fleet. But then Jellicoe, in spite of major blunders by Admiral Reinhard Scheer, and in spite (or because) of executive signals at a rate of 1 every 67 seconds, failed to get inside the Germans' action–reaction cycle, with the result that they escaped, not once, not twice, but (if you count the events of the night) three times, to get home first and claim victory.[2]

The respective standing orders of Jellicoe and Beatty reflected their personal differences. Whereas the keynotes of Jellicoe's Grand Fleet Battle Orders were detailed planning, command centralization and a dependency on signals from the flagship, Beatty's planning for the battlecruisers was much less constrained, relying on initiative (or command by negation) and the minimum of signals from the flagship. It was the latter that seemed to work, and, in the aftermath of the war, it seemed impossible for such empirical lessons to be ignored. With the Beatty faction dominating the post-war navy, a change of doctrine seemed to be both obvious and inevitable.

At the end of the Great War, the Royal Navy began scrutinizing its performance, asking questions, and looking for lessons about the types of ship, and their use in a future war at sea. The Royal Navy had not fought the decisive battle against the High Seas Fleet that many, within both the navy and the general population, had been led to expect; Jutland, the major surface action between the British and German battle fleets, had not been decisive and the German surface fleet had remained a potential threat. Strategically, the Royal Navy had restricted the use of the German High Seas Fleet, which had, in its turn, become a 'Fleet in Being', but whilst the Grand Fleet swung

INTRODUCTION

on their buoys at Scapa Flow, the German U-Boats had nearly starved Britain into submission in 1917. Faced with this fact, and the post-war decline in the interest of anti-submarine warfare in favour of the battleship, the concentration on the decisive battle, often dismissed as re-fighting the Battle of Jutland, has been portrayed in an unfavourable light, clouding the Royal Navy's thinking on the proper course of future developments, with near disastrous consequences in the Second World War.[3]

Therefore, in planning for a war against Japan, War Memorandum (Eastern) provided the focus for the development of a balanced battle fleet during the 1920s and 1930s and also helped to shape the new tactical developments that were seen as necessary for the smaller British battle fleet to obtain victory against an enemy. And the earnest desire *not* to re-fight Jutland did produce improvements in gunnery, communications between ships, the development of offensive divisional tactics, the conduct of night actions, and more offensive use of destroyers, as well as contributing to the development of a doctrine for the use of aircraft at sea. These were all increasingly reflected in the Battle Instructions, as well as in the annual Fleet Exercises and numerous Staff College Exercises. However, the battle fleet remained central to naval thinking, and, in the immediate post-war period, a major debate raged over whether battleships should be the capital ships of the future, or whether more submarines (including the semi-submersible, or even submersible battleship), aircraft carriers, and their own or land-based aircraft, could be a realistic substitute in any future navy.[4]

Such decisions rested on how the navy might be used in the future, and which country (or countries) would be a future enemy, and these issues formed the basis of both formal and informal debate within the Royal Navy. One such possible enemy was the United States. Affronted by their inability to assert the rights of neutrals in wartime, they had, since 1916, been pursuing a building programme with the widely known aim of producing a fleet that would successfully challenge the Royal Navy's traditional maritime supremacy. With the previous challenger to Britain's naval supremacy, Germany, defeated in a war, the Royal Navy still possessed the world's largest battle fleet, but both the United States and Japan had naval building programmes whose ships threatened to outclass the Royal Navy; this problem was compounded by the dilemma of whether to have the United States as an ally or renew the Anglo-Japanese Alliance and alienate the United States. However, not only did professional links (most recently demonstrated in the Great War) and ties of kinship seem to make such a war unlikely but, pragmatically, Great Britain could not afford either another naval race or a war with the United

States, and a paper, written by Admiral Sir William Lowther Grant, on striking his flag as Commander-in-Chief, North America and West Indies Station, made the point that the Americans would settle for parity with the Royal Navy, especially if the British abandoned their alliance with the Japanese. Grant also felt that the British had made no attempt to explain their unique, imperial situation to the Americans and that doing this was vital, to avoid tensions. They had to make clear to the Americans that a large Royal Navy was essential to fulfil their imperial obligations, which could not be left to any other power, such as Japan, to safeguard. Grant advised that the British

> fix and publish the minimum strength we adopt as a policy and to which we intend to build up to or gradually reduce, at the same time make it clear that this is done regardless of US, whose present programme is not taken seriously. Our standard will undoubtedly be adopted by the US as their standard.[5]

What Grant proposed was what successive British governments were unwilling to do during the 1920s and 1930s: determine future imperial naval requirements, let the Americans know what these requirements were, and why they were necessary, and, as the Americans were extremely unlikely actually to build up their fleet to the suggested size, leave them to concentrate on building a fleet to meet their own needs, not to compete with the Royal Navy; in short, have parity for the Royal Navy and United States Navy battle fleets.

Lloyd George, the British Prime Minister, believed that given the choice between an alliance with Japan or an agreement with the United States, Britain would be better served by having the United States as an ally. Not only would possible clashes be avoided, with both the United States and the Dominions, but an alliance or agreement with the United States could help to bring it back into the international diplomatic field, and away from its increasingly isolationist policies.

This left Japan as the only 'potential enemy' – the only power possessing a fleet and a building programme large enough to threaten Great Britain's supremacy at sea – and a large British battle fleet was seen as vital to guard against any future Japanese aggression. After all, conditions in the Pacific were different from those in home waters, with long sea lines of communications thousands of miles long separating the Dominions' widely scattered populations and under-exploited resources. Only by considering the strengths of the opposing fleets, the geopolitical features and the lines of communications could an appropriate strategic plan be built up and naval strategy developed.

Japan, an ally and the third naval power in the world, appeared set on expanding its influence and importance in the Far East. Faced with increased US and British building programmes, Japan wanted to continue its own '8:8' building programme, eight modern battleships and eight modern battlecruisers, supported by the existing fleet and which both the British and Americans would have to match to avoid being left behind.

Although this seemed common sense, not everyone agreed. In many ways, the Japanese challenge was seen to be the greater threat to British power and prestige in China, the Far East and the Pacific and although the continuation of the Anglo-Japanese Alliance was seen as a way of keeping Japan in check, this was causing serious disquiet in Australia, New Zealand and India. In 1920 the Foreign Office decided that whereas the interests of Great Britain and the United States were often similar with regard to China, they were increasingly clashing with those of the Japanese, who were taking a lead in an 'Asia for the Asians' movement and encouraging nationalism in India and in China.[6]

It appeared far more likely that Japan would fight the United States, leaving a strong British fleet to act as a moderator, and therefore no special preparations needed to be made against the Japanese. In fact, the likelihood of Japan going to war with Britain was seen as so remote that Lloyd George, in an address to the Imperial Conference on 20 June 1921, stated his belief that Japan would not abandon its traditional policy by sending its fleet thousands of miles from secure, home waters to engage an enemy battle fleet. This was an argument also used by Churchill, when Chancellor of the Exchequer, to push for a reduction of the Naval Estimates and one which Beatty and his successors as First Sea Lord tried to resist, with varying degrees of success.

Thus the idea of a war with Japan came to dominate Admiralty thinking, both as an end in itself and as a means of maintaining the largest possible Naval Estimates. War with Japan was not a new idea, having been mooted first by the British as far back as 1909, when the Imperial Conference had agreed that whilst Japan remained the guarantor of British security in the Pacific Ocean, a significant imperial naval presence was still necessary, as sending naval forces eastwards in a period of tension might actually provoke Japan into declaring a war. So, after the Great War, when the Australian Prime Minister, William Hughes, spoke of the political and strategic shift in emphasis away from the European and Mediterranean stage towards the Pacific, he was voicing a long-held opinion, both in the Dominions, and within the Royal Navy.[7]

From as early as October 1919, the whole question of a naval base at Singapore was to become a central issue, as the Naval Staff questioned the government regarding policy towards the Far East. If the alliance with Japan was to be terminated in 1922, then the Admiralty believed that there was a need for a major naval base in the region and a fleet, including Dominion contributions, capable of operating from such a base. Jellicoe's Naval Mission to the Dominions, which had followed their rejection of an Imperial Fleet, was in response to suspicions of Japan's intentions in the region, although in some ways Jellicoe identified more problems than he offered solutions, and his mission developed in a way never envisioned by the government. He increasingly began to think that Great Britain and Japan would inevitably clash, and that without adequate naval preparations the Royal Navy would be dangerously divided, with the bulk of the battle fleet in the Far East, leaving the United States Navy supreme in the Atlantic.[8] If the United States intervened to protect its neutral rights during a war, when most of the Royal Navy was attempting to stop Japanese trade with the United States, another powerful fleet would have to be found and sent to disrupt US trade, and this would inevitably delay sending the full fleet to the Far East. In the meantime, US Asiatic forces would be able to cause considerable disruption to British trade in the China Seas. And if the Main Fleet was in the Far East, it was conceivably possible for the United States to blockade the British coast. Recognizing this fact was the start of a gradual recognition that the Royal Navy could no longer operate in two hemispheres without risk. His solution, two battle fleets, one partly financed by the Dominions, was unacceptable both to the Dominion governments and to the Admiralty, on the grounds of expense. Also, the Dominions needed to know who would be in control of the vessels. But if the British could no longer afford to have a fleet twice as large as its rivals, then its sea power alone could not hope to defeat decisively a major non-European naval power, and either political decisions would have to be made as to where to abandon to the enemy, or war would have to be avoided at all costs.

After the signing of the Washington Naval Treaty, however, the Royal Navy achieved parity with the United States Navy in capital ships, and superiority over the Imperial Japanese Navy, by slashing existing construction programmes (largely on paper in the British case, although many of the older British battleships also had to be scrapped) and limiting future construction, until after a ten-year 'holiday' on building.[9] The Royal Navy was left with a smaller battle fleet than before, but one which was still numerically and qualitatively superior to both the United States Navy and the Imperial Japanese Navy, and which would, by virtue of the moratorium on building, remain so until the early 1930s.

Reactions to the Washington Treaty differed, both at the time and subsequently, and the negative reaction of many has formed the basis of the view that the Royal Navy suffered severe cutbacks as a result. This was not actually so. The Admiralty saw this as a period in which they could plan and construct the necessary ships for a war with Japan, which would not, it was thought, start until some time after 1931. Government ministers, by contrast, saw it as a removal of a threat from Japan for at least a generation, presenting an opportunity to cut back on the annual Naval Estimates.[10] To the government, the battle fleet did not need to be ready by 1931, and if there was a threat in the future, there would be ample warning and preparation time before a war broke out. This difference of opinion was largely the reason for the struggles between the Cabinet and Treasury, on the one hand, trying to curb the Services' spending, and, on the other, the Admiralty, which pressed for the necessary money to develop the fleet.

There were still doubts over the Japanese threat, and the Royal Navy's ability to counter it, in the Dominions, and they refused to accept the Admiralty's notions of an Imperial Fleet financed and manned by the Empire but centrally controlled by the Admiralty. In 1923 the Labour Minister of Defence in Hughes' government put the politically sensitive view to the Australian Senate that Australia would be better off, and more secure, if they spent money on aircraft for the defence of Australia, rather than on battleships and their related auxiliaries, thereby depending on the Royal Navy, for protection.[11] Many others in Australia hoped to create closer links with the United States so that in time of war the US Pacific Fleet would be deployed to bases in Australia, and saw the cruise of the US Pacific Fleet to Australia as proof that this was possible.[12] Either action could have spelt the beginning of the end for British imperial control in the region, but although keen to keep the Australians closely tied in to the overall strategic plan, the Admiralty rejected any ideas of developing major naval bases in Australia because they felt that this would send the wrong signals and reverse priorities, making the defence of other regional possessions dependent on the defence of Australia.[13]

All of this meant that imperial defence remained a major problem for British governments after the Washington Conference, and that

> The challenge was to provide an adequate system for protecting the far flung British Empire, including its world wide network of sea lanes,

while dealing with a static situation in regard to constraints, as well as resisting a potential naval race on cruisers and submarines. Yet the Admiralty's own emphasis on capital ships diverted the Royal Navy's effectiveness from protection of the trade routes and tactical predominance in the Atlantic theatre to a preoccupation with the impossible task of fleeting-up distant strongholds like Singapore.[14]

There was, however, no clear consideration of world-wide strategic issues by the British government until the mid-1930s, with the deterioration of European security. Despite this, the Washington Treaty had left the Royal Navy with a modern battle fleet, equal to the US fleet, superior to the Japanese and believed to be qualitatively superior to either of these navies. Without the Washington Treaty, the Royal Navy would actually have been in a worse situation, with an ageing battle fleet and the likelihood of a naval race, if not outright war, with the United States. The Treaty gave Britain parity in capital ships and by shifting the focus on to cruisers gave Britain another advantage over the United States and Japan, as at the time it had an overwhelming advantage in modern cruisers.[15] By signing up to the Washington Conference, Britain had saved itself millions of pounds on constructing ships and the Royal Navy remained superior to any other of the major navies, with sufficient superiority to consider sending a fleet to the east, as well as having vessels to act as a deterrent in European waters.[16] What the Royal Navy really needed in order to do this was a war plan to show how it intended to maintain security and superiority: the result was War Memorandum (Eastern).[17] At first, this was only a plan for moving ships eastwards in time of war, but it gradually developed into a more elaborate plan. Raids by cruisers and aircraft carriers were all aimed at forcing the Japanese battle fleet to abandon its own strategy and to seek out the British battle fleet for a decisive battle at sea.[18] So it was that the Japanese battle fleet, and not the US battle fleet, became the justification for the Admiralty's construction programmes, and the measure of the Royal Navy's efficiency in time of war. European waters were still an important area of operations, with possible threats from the French and later the Italian fleets, across the lines of communication through the Mediterranean Sea from Gibraltar to the Suez Canal and thence to the Far East, but with the major portion of the battle fleet based in the Mediterranean, able to be moved eastwards or westwards as circumstances dictated, it was believed that these could be neutralized. And there was also the alternative but longer route around the Cape of Good Hope.

Consequently, it was the problems associated in moving, maintaining and operating a battle fleet in the Far East that began to preoccupy the Admiralty throughout the 1920s. For the first time in its history the Royal Navy was planning to deploy its major ships thousands of miles from home waters, and to keep them fully supplied, repaired and in fighting order. The establishment and development of a major naval base at Singapore, able to defend itself until the arrival of the Main Fleet from the Mediterranean, and the strategy of maintaining a fleet in the Far East became a major priority and it is no understatement that

> [t]he barometer of British naval policy in the 1920s with regard to Japan was provided by the rate of progress of the Singapore naval base because without a fully equipped and defended base in the Far East, Britain would be unable to dispatch her battle fleet from home waters and the Mediterranean, and would therefore be powerless to protect her interests and territories, or even exert any political leverage in the Far East.[19]

The British strategic objectives were primarily defensive, protecting British possessions and trade by a tactical offensive, although there was, through the 1924 War Memorandum (Eastern), the belief that Japan, unlike the United States, could be isolated by a vigorous offensive against its sea-borne trade. The destruction, or at least the containment, of the Imperial Japanese Navy's battle fleet would greatly assist in this by removing this threat to British cruisers. But, as both a contemporary critic of War Memorandum (Eastern), Captain Herbert Richmond, and a later First Sea Lord, Admiral Sir Roger Backhouse, pointed out, there were many ways of making war at sea without fighting a fleet action, as the recent maritime war against Germany had so graphically illustrated. Subsequent British war plans were framed on the assumption that the Japanese would not voluntarily fight a fleet action away from its home waters, especially if there was a chance of it losing and that the Royal Navy would have to be prepared to launch attacks against Japanese light forces to compel the Japanese battle fleet to sortie and defend these.

In 1921 the Royal Navy was still a large and potent force, theoretically able to send a large battle fleet to the Far East. The Atlantic and Mediterranean Fleets were the major naval commands, whilst less senior flag officers led cruiser squadrons on the North American and West Indies, China, East Indies and African stations. There were also squadrons of Australian, New Zealand and Canadian ships, local defence and training flotillas and a sizeable Reserve Fleet. Critics of the emphasis on the surface, fleet action, such as Richmond, believed that if the navy was only organized for a battle fleet action, this would be all it was good for, and that the preservation of such an attitude would repeat the mistakes of the last war, ignoring trade protection in favour of preserving a large battle fleet ready for decisive action.[20] It was partly because of this that Richmond continued to call for greater consideration of strategic priorities and smaller, less expensive battleships, whilst the Admiralty ignored him, holding to the belief in the decisive fleet action, and that having the heaviest armed and armoured battleship possible, which could have a prolonged life through refits and modernization, was the more sensible and economic alternative.

The emerging Japanese threat was the only concrete justification for an up-to-date battle fleet that Beatty and his colleagues could use when framing the Naval Estimates, and the strong inter-Service rivalries for the largest Estimates possible, combined with the initial lack of a clear government strategy, exacerbated this. Government was declining to proffer any strategic planning until it was reasonably certain about what the post-war world order would be, and it assumed, quite reasonably, that the League of Nations would be acting as a coalition of powers to underwrite any agreements. But the Cabinet did not control the Admiralty's policy-making and nor could the Treasury, without their support. They could inhibit it, and Cabinet could prescribe what *not* to do, but, until a definite policy was formulated, all three Services could, and did, develop their own, ambitious plans and ideas.[21]

However, there were objections to the Admiralty's policies and approach to strategic planning, predominantly on the grounds of cost. As a Cabinet member and Chancellor of the Exchequer, Winston Churchill objected to the Admiralty challenging Cabinet decisions, and, as had Lloyd George, countered their claims regarding the Japanese threat by ingeniously suggesting that if no preparations were made to fight the Japanese in the Far East, the task of protecting British interests would become easier. He argued thus:

> The Japanese Fleet cannot come around the world to attack our fleet in home waters or in the Atlantic Ocean, and they would place themselves at an enormous disadvantage if they attempted to fight a decisive fleet action in the Indian Ocean. Therefore, a decisive fleet action cannot take place unless we ourselves send our fleet around the world into the Pacific... no fleet action can take place unless both sides wish it.[22]

Whilst much of what Churchill said was true – the Japanese would also have had difficulties in operating thousands of miles from home waters, and it would be impossible to fight unless both sides wished it – he chose to ignore the deterrent effect that the Admiralty claimed could be exerted on the Japanese. Making no preparations could surrender control of the western Pacific to Japan, expose Australia and India to raids and undermine both Britain's commerce and prestige in the area. Re-exerting any influence in wartime would also have been much more difficult. The Japanese would not want to fight on Britain's terms, and at a time of Britain's choosing, and consequently any war would inevitably be a long-drawn-out and difficult one. So when the Admiralty was told what it could not plan for, war with Japan, they simply made a professional decision and continued to formulate naval policy and procurement with a future war in the Far East in mind. Naval policy was traditionally to see every naval power as a potential threat to imperial security, requiring a large, balanced fleet, but by focusing on one threat, Japan, and on the surface battle, the Admiralty believed that they were creating a navy that was not just capable of carrying out a Far Eastern war plan, but one which would be able to cope with most other contingencies.[23]

In the event, the Cabinet managed to exert its primacy and judged that war with Japan was not likely and that the One Power Standard was to be redefined to mean that the Royal Navy was to be equal to the battle fleet of any potential enemy; local superiority was not required, and local naval forces could be inferior in strength until the arrival of the main battle fleet.

In the 1930s, expediency also played a major part in this change in emphasis as the ageing battle fleet, with its decreasing ability to operate from forward bases such as Hong Kong, forced the planners to consider a more defensive strategy. Plans were also increasingly complicated by the need to keep a significant number of vessels in European waters to match the Italian and the growing German battle fleets.

At the operational and tactical levels this did not seem to impinge greatly on the institutional self-image of the Royal Navy. This self-confidence stemmed from the generally held conviction that not only did sea power remain a potent weapon of war which could attain decisive results, but that British material and superior training would produce decisive results against Japan. In the 1920s the Royal Navy remained confident in its ability to project power into the Far East and defeat the Japanese, but, until 1931, when everything was expected to be in place, the development of the Singapore base, and hence of the entire Far Eastern policy, was all a huge gamble. It was based on the hopeful assumption that the Japanese would not be too aggressive, or at least, that if they were, they would not be too aggressive too quickly, and would allow the British the 30–60 days they expected to need to send their battle fleet eastwards.

Tactically, developments were also taking place and the Royal Navy's Battle Instructions, offered guidance as to how a smaller, post-Washington fleet would fight, benefiting from the lessons of Jutland. These naturally laid a great deal of emphasis on the use of the battleships and their supporting ships, but also stressed the growing importance of the aircraft carrier and its aircraft in future fleet actions. Many, within both the Royal Navy and the Royal Air Force, advocated the use of air power and claimed that aircraft had made the battleship obsolete, and in order to counter this the Admiralty devoted considerable time and effort to showing that naval aviation could not seriously replace the battleship, but that it could be of great assistance in helping the battleship achieve its objective, destruction of the enemy. Far from being conservative, big-gun and 'anti-aircraft', the Admiralty planners saw a whole range of roles for aircraft in a fleet battle, and wrote these into the Battle Instructions. Aircraft were to be an integral part of the battle fleet, both before and during an action.[24] The Royal Navy was committed to taking aircraft to sea and using them and it was other, political, constraints that hindered their fullest exploitation, most notably in the case of aircraft development and procurement, although it is important to recognize that the RAF was not the 'villain of the piece', as is sometimes portrayed; the Air Ministry generally provided the type of aircraft that the Admiralty asked for.[25] The Royal Navy also possessed the largest aircraft carrier fleet during this period but it was the lack of expertise in naval aviation – once many of the fliers had transferred to the RAF – as much as the lack of aircraft that dictated that they be used closely with the surface fleet (instead of developing massed, offensive air tactics) rather than conservatism or a lack of enthusiasm for their use. Such a doctrine stands in sharp contrast to the way that the United States Navy and the Imperial Japanese Navy, both of which retained control of aircraft procurement and deployment, were able to develop their naval air arms. But even in these navies, the battleship was still seen as the deciding factor in naval warfare, and it was only the lack of them, after Pearl Harbor, which prompted the Americans to rely on the carrier task group. The Japanese, by comparison, continued to look for the decisive gun battle until quite late on in the war.

The Royal Navy had thus developed a war plan, involving a battle fleet deployed thousands of miles from home waters, able to rely on a major naval dockyard, Singapore, for supplies, repair facilities and ammunition, in addition to a supply chain, secure lines of communication and military support from the Dominions. Tactically, this suggested fighting as quick and decisive a war as possible, and forcing a decisive action at the earliest opportunity. A slow, measured build up, and a prolonged war against Japanese trade would work against the British ships. Attrition, and mechanical breakdown would wear them down. Supplying them would strain the mercantile resources of the Empire and could not be afforded.

Therefore operational plans and tactics had to be developed which reflected this and catered for a quick war with a decisive battle fleet action, fought as early as possible. This would also place Japanese forces on the defensive and remove the threat of Japanese battle-cruisers acting against British cruisers attacking Japanese trade, which would, it was hoped, prompt the surrender of Japan. Such a war had to be a naval war, partly due to the nature of the region, and partly because land operations in China, outside of the defence or recapture of Hong Kong, or a raid on outlying Japanese territory, was deemed to be impossible.

The Admiralty continually presented its case for a navy sufficiently large to ensure the safety of British possessions and maintain sea lines of communications. Their reasoning though was often too imprecise to politicians keen to accept naval arms limitation and seeking cuts in expenditure, despite the erosion of the pool of specialist warship builders and the loss of skilled men from their yards that this would entail. The danger was that at the end of the ten-year building holiday, with an ageing British fleet needing to be modernized and replaced, the shipyard capacity and workforce to do the jobs would no longer be there.

Expenditure on the Royal Navy, and financing its plans at a time of economic depression continued to be a bone of contention, and in 1923, on the heels of the 1921 'Geddes Axe', a committee, chaired by Sir Alan Anderson, reported on naval pay. The committee concluded that naval pay was too high relative to other workers and in 1925 an Admiralty Fleet Order established a lower rate of pay for men joining after 5 October, but seemingly guaranteeing the older, higher rate to those who had joined before this date.[26] After the 1929 Wall Street Crash and the start of the Great Depression in Britain in 1931, and with the country in a grave financial state, another committee was set up, chaired by Sir George May, to determine how to cut public expenditure even further and retain the Gold Standard. His committee decided that cuts in naval pay would be necessary and that no guarantees regarding the higher rate of pay had been made. Despite the First Sea Lord's protests, the First Lord, Alexander, had little alternative but to announce that the Royal Navy would accept the May Committee's recommendations.

Most of the Royal Navy's personnel found this out from the press and not from their officers. The result was the already well-documented Invergordon Mutiny. Summer leave had already been taken by many of the men of the Atlantic Fleet and they had had time to reflect on the effects of possible pay cuts of up to 25 per cent for an able seaman. The timing of the announcement to the fleet was mishandled, and in September 1931, the crews of the *Nelson, Rodney, Hood, Valiant, Norfolk* and *Adventure*, ships of the Atlantic Fleet at Invergordon, refused to put to sea for exercises. The fleet had to be dispersed and returned to its home ports. This was a low point for morale in the Royal Navy and great damage was done, not just to the prestige and image of the Royal Navy, but also to the nation as a whole; confidence in Britain slumped, inflation rose and the government had to abandon the Gold Standard.[27] For the rest of the decade the Royal Navy would struggle, not just to build up the battle fleet at a time of economic constraint, but also to rebuild its self-confidence.

Throughout all of this, planning for a war against Japan, and for the provision of a naval base at Singapore continued. The Admiralty frequently had to argue in Cabinet that it, and not the Treasury, should define Britain's naval and maritime needs. Nevertheless, there is no doubt that the Cabinet, whilst forbidding the Admiralty from preparing plans for offensive operations against Japan, expected the Royal Navy to be able to complete its preparations for safeguarding imperial maritime security, which included providing a base at Singapore, reserves of fuel, and replacement and construction programmes. In this, the Admiralty was reasonably successful. Large numbers of war-weary ships had been scrapped, a Reserve Fleet was maintained, new ships, especially cruisers, destroyers and submarines, were designed and built, new equipment, such as the Admiralty Fire Control Table, improved boilers and catapults were procured, and programmes for new construction were maintained, subject to Treaty limitations, preserving the manufacturing base as much as was possible given world economic trends. So, in maintaining a balanced battle fleet and focusing on an extreme case, war with Japan, the Admiralty was attempting to maintain a fleet capable of a world-wide deployment, and concentrating on a battle fleet, with the battleship at its core, supported by aircraft carriers, cruisers, destroyers and submarines. Despite the widely held belief that the submarine had been largely mastered by ASDIC (named after the Anti-Submarine Detection Investigation Committee), battleships were still expected to be screened by escorts and 'fleet submarines' appeared in the Battle Instructions, even though such vessels did not exist and the Tactical Division did not really see a need for them, preferring instead an effective patrol submarine which could reconnoitre when the weather was too poor for aircraft to operate, or when surface forces would be too vulnerable. The belief that submarine warfare against commerce would not be a major feature of a future war limited, but did not completely stop, the development of escort vessels which could be acquired easily and built in wartime, and prototypes were designed and built.[28]

The battle fleet that was expected to face a Japanese fleet in battle may certainly have been smaller than that at Jutland, but with the use of smaller, more cohesive formations, concentrated fire, aircraft and destroyer attacks, it was expected to be much more aggressive. It would also be more independent: its flag officers were expected to use greater initiative and be confident of victory. Developing this independence of action was going to be just as vital as providing the ships, but measuring success was going to be difficult. According to Dewar, writing in *The Navy from Within*, the Royal Navy's culture was against such an innovation, and he wrote tellingly that, 'In the British Navy ... capacity to think was a handicap and an independent or critical mind was a definite disability ... serious thinking may be more dangerous than serious drinking.'[29] Similarly, Arthur Marder was able to find several disgruntled senior officers all of whom complained that 'Enterprise, initiative, enthusiasm, inventiveness, departure from the official mould, and a host of valuable attributes were all discouraged if not regarded as objectionable; even individuality or a too enquiring mind were out of place'.[30] More recently, it has also been stated that the Royal Navy's approach to training its young officers was to force a young and impressionable mind into a standard mould, so that what emerged was a fit, tough, highly trained but sketchily educated professional, 'ready for instant duty ... for peace or war; confident leaders, alert seamen ... officers of wide interests and narrow vision, strong on tactics, weak on strategy'.[31] The unspoken assumption remained that a good, practical, sea-going lieutenant would make a good, practical, sea-going admiral, and no attention was devoted to encouraging officers to think about their profession.[32]

Attempts to address this problem met with only limited success. Marder commented on how it was curious, but not surprising, to see how badly officers expressed themselves, compared to the soldiers on the course, when at the War College, revived in 1919, and how it was therefore unsurprising that naval officers, up to and including Sea Lords, were unable to stand up to politicians. Beatty only managed it by having a forceful personality, *and* by threatening a mass resignation by himself and his Board, whilst Madden and Field both have been unjustly judged as caving in to political pressure during their tenure of office. (In their defence, Madden can be credited with ensuring that the new cruisers built after the 1930 London Conference were modern enough to be deployed to Far

Eastern waters, and Field was far-sighted enough to appoint Henderson as the first ever Rear Admiral Aircraft Carriers.)

It has been suggested recently that the problem was thrown into high relief by the new requirement for staff work. Although the Royal Navy emerged from the Great War convinced of the value of a Naval Staff, it has been claimed that little was done to supply it with the calibre of officer necessary, that the best officers went to sea and the Naval Staff got what was left.[33] However, the Naval Staff did also get Chatfield, Fisher, Backhouse, Cunningham and Pound, all significant figures for the Royal Navy in the years immediately before, and during, the Second World War.

The greatest achievement of these and others was the change around in the Royal Navy's self-image. Instead of being burdened with the failure to achieve a decisive battle against the High Seas Fleet at Jutland, under Beatty's leadership the Royal Navy assumed another, quite different, mantle, that of decisive battle and ultimate victory. Even though the eventual war against Japan was not fought in the way envisaged by the strategists and changing circumstances made it impossible to undertake War Memorandum (Eastern), the tacticians of the 1920s and 1930s succeeded in reinstilling aggressiveness and confidence into the Royal Navy. Their efforts did bear fruit time and again, when naval officers of all ranks and backgrounds had to face the enemy and make instant decisions. The victories of the Battle of the River Plate, the Battles of Sirte and Matapan, the sinking of the *Bismarck* and the sinking of the *Haguro* are but a few examples, showing how much the Royal Navy had absorbed the tactical principles implicit in the Battle Instructions of the period, which in their turn had been partially influenced by the demands of fighting a naval war against Japan.

NOTES

1. G.A.H. Gordon, *The Rules of the Game: Jutland and British Naval Command* (John Murray, London, 1996).
2. G.A.H. Gordon, 'The Doctrine Pendulum', in P. Hore (ed.), *The Hudson Papers*, Vol. I (Ministry of Defence ([Navy], 2001), p. 95.
3. The standard work on the period is Captain S.W. Roskill's two-volume work, *British Naval Policy between the Wars* (Collins, London, 1968). Roskill comments on what he believes is the obsession of Jutland on p. 533. Similar views are expressed by Correlli Barnett in *Engage the Enemy More Closely: The Royal Navy in the Second World War* (Norton, New York, 1991). He captions a photograph, between pp. 82 and 83, thus: 'Before the Second World War ... a future Jutland against the Japanese fleet in Far Eastern waters befogged the Royal Navy's thinking.'

INTRODUCTION

4. See, for example, PRO, ADM 116/2060, 'Post War Questions Committee, 1920–21'.
5. M. Simpson (ed.), *Anglo-American Naval Relations 1917–1919* (Navy Records Society/Scolar Press, Aldershot 1991), pp. 593–4.
6. W.R. Louis, *British Strategy in the Far East 1919–1939* (Clarendon, Oxford 1971), discusses British policy with regard to China.
7. P. Sales, 'The Naval Mission and Australian Security after the First World War', in D. Stevens (ed.), *Maritime Power in the 20th Century: The Australian Experience* (St Leonards, New South Wales, 1998), p. 40.
8. For the Jellicoe Mission see A. Temple Patterson (ed.), *The Jellicoe Papers, Vol. II* (Navy Records Society, London, 1968), pp. 284–397; N. Tracy (ed.), *The Collective Naval Defence of the Empire 1900–1940* (Navy Records Society, London, 1997), pp. 241–74.
9. J. Ferris, '"It Is Our Business in the Navy to Command the Seas": The Last Decade of British Maritime Supremacy, 1919–1929', in K. Neilson and G. Kennedy (eds), *Far Flung Lines: Studies in Imperial Defence in Honour of Donald Mackenzie Schurman* (Frank Cass, London, 1997), pp. 133–4.
10. Churchill wrote to Admiral Sir Roger Keyes, in 1925, to the effect that there was no danger of Japan attacking the British Empire and that 'The great task before the Admiralty is to try to rebuild the Fleet in a sober and reasonable way, and without injuring the general interest to an excessive extent. To do this, they must: a) drop the idea of a war with Japan; b) abandon the plan of putting a Fleet in the Pacific; c) reduce the expense and maintenance of the fleet', P. Halpern (ed.), *The Keyes Papers, Vol. II, 1919–1939* (Navy Records Society, London, 1980), pp. 110–11.
11. G.H. Gill, *Royal Australian Navy*, Vol. II (Tollins, Sydney, 1957), pp. 17–18.
12. Prime Minister Bruce, meeting with Australian Service chiefs before the 1923 Imperial Conference, said that, 'Although the Singapore project was an indication that Great Britain had recognized that the Pacific had become an area of possible change, it did not follow that the whole strength of the British Fleet would be stationed in Eastern waters'. At the same meeting, Captain G.F. Hyde, RAN, pointed out that, without Singapore, the United States could offer no help in the western Pacific, but that with such a base, they could become a powerful ally. Sales, 'The Naval Mission', p. 47.
13. The 1921 Commanders-in-Chief Conference, held on board HMS *Hawkins* at Penang in March 1921, concluded that Singapore was the only suitable site for a main fleet base. Tracy, *Collective Naval Defence*, pp. 278–9.
14. Sales 'The Naval Mission', p. 45.
15. Ferris, 'It Is Our Business', p. 133.
16. Ibid., pp. 133–4.
17. PRO, ADM 116/3125, 'War Memorandum (Eastern), August 1924'.
18. J.R. Hill (ed.), *The Oxford Illustrated History of the Royal Navy* (BCA, London, 1995), p. 346.
19. Barnett, *Engage the Enemy*, p. 23.
20. D. Baugh, 'Admiral Sir Herbert Richmond and the Objects of Sea Power', in J. Goldrick and J. Hattendorf (eds), *Mahan is Not Enough – the Proceedings of a Conference on the Work of Sir Julian Corbett and Sir Herbert Richmond*, pp. 24–5; C. Hill, '"How Are We Going to Make War?": Admiral Sir Herbert Richmond and British Far Eastern War Plans', *Journal of Strategic Studies*, Vol. 20, No. 3, September 1997, pp. 123–41.
21. J.R. Ferris, *The Evolution of British Strategic Policy 1919–1926* (Macmillan, London, 1989), p. 54.

22. Ferris, *British Strategic Policy*, p. 164.
23. Ibid., p. 80.
24. PRO, ADM 186/72, 'Battle Instructions 1922–1927'; G. Till, *Air Power and the Royal Navy* (Janes Publishing, London, 1979), pp. 138–42.
25. D. Hobbs, Curator, Fleet Air Arm Museum, Yeovilton, to author, 21 May 2001.
26. Captain John Wells, RH, *The Royal Navy: An Illustrated Social History, 1870–1982* (Royal Naval Museum/Alan Sutton, Stroud, 1999), pp. 141–5.
27. Ibid.
28. Remarks by D. Baugh, 'Confounded by Perplexities: The Navy and British Defence Planning Between the Wars', Conference on 'The Parameters of Naval Power in the Twentieth Century', University of Exeter, 4–7 July 1994.
29. N.A.M. Rodger, 'Training or Education: A Naval Dilemma over Three Centuries', in P. Hore (ed.), *The Hudson Papers, Vol. I*, p. 15.
30. A. Marder, *Old Friends, New Enemies: The Royal Navy and the Imperial Japanese Navy, Vol. I: Strategic Illusions, 1936–1941* (Oxford University Press, Oxford, 1981), pp. 281–2; Rodger, 'Training or Education', in Hore (ed.), *Hudson Papers, Vol. I*, p. 18.
31. Rodger, 'Training or Education', in Hore (ed.), *Hudson Papers, Vol. I*, p. 24.
32. Ibid.
33. Ibid., p. 23; Admiral King Hall, *My Naval Life, 1906–1929* (Faber & Faber, London, 1952), p. 215, describes naval staff officers as 'dead from the neck up'.

1

The Influence of a Far Eastern Strategy on British Naval Policy

Japan took its commitment to the Anglo-Japanese Alliance seriously in the Great War, participating in the hunt for Graf Spee's squadron, escorting convoys, seizing German colonies in China and the Pacific and even putting down a mutiny in Singapore. Japanese shipyards also built ships for the allies, including a class of 12 destroyers for the French Navy. But Japanese national pride was affronted by British requests in 1916 either to buy their modern *Kongo*-class battle cruisers, or to have them manned by British crews and deployed with the Grand Fleet, to replace war losses. Not only did the Imperial Japanese Navy see the manning of Japanese ships by British crews as a national slur and an insult on the efficiency of their own navy, but they also saw the loss of these vessels, even temporarily, as a setback to Japanese naval expansion, based on the 8:8 Plan to provide a powerful battle fleet to equal US naval power. However, in 1917 the Japanese agreed to deploy cruisers and destroyers in the Mediterranean to protect convoys, and there the British were impressed, and had had nothing but praise for Japanese efforts and efficiency.[1]

The employment of Japanese ships in the Pacific had provoked a mixed response in Australia and New Zealand though. Whilst grateful for the Japanese contribution, the Dominion governments were fearful of further Japanese expansion towards their territories and disapproved of the seizure of German possessions in the Marshall, Marianas and Gilbert Islands, and especially the important cable station on Yap Island.[2] Consequently, after 1918 Britain and Japan moved further apart from each other and, in April 1919, the Director of Local Defences Division (DLDD), Commander Larken, noted:

> The strategical centre of gravity may be said to have shifted from the North Sea to the Pacific, and future Naval policy depends on our

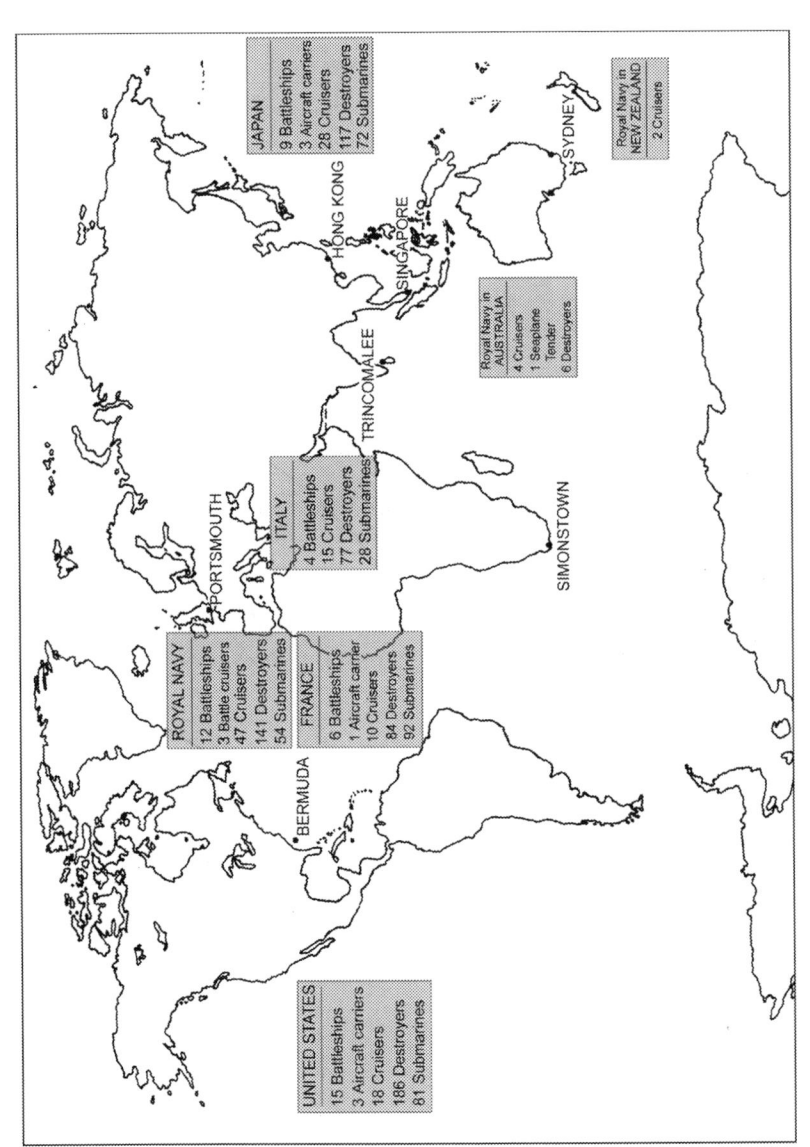

Map 1: The Major Battle Fleet Strength of the World, 1931

relations with Japan. The sooner therefore, that the intentions of our government are known in regard to the renewal, or otherwise, of the Anglo-Japanese Alliance, the better.[3]

He was expressing widely held beliefs; the Pacific was now regarded as the major strategic area of importance, and the British government needed to provide guidance regarding the attitude to be adopted towards Japan. However, British politicians were unwilling to make any hard and fast decisions until the future shape of the post-war world had finally been resolved by all of the various treaties and negotiations, even though political leaders such as MacDonald, Baldwin and Churchill acknowledged that Britain's Empire, trade and status all depended on the Royal Navy. But the Royal Navy's views on what was regarded as necessary to maintain command at sea – a large and well-equipped battle fleet – seldom accorded with what the Treasury thought they should be spending, and the definition of strategy and sea power often came to be decided by how much money the Treasury was prepared to grant the Admiralty, rather than being based on strategic priorities.[4]

War Memorandum (Eastern) and the development of the Singapore naval base are prime examples of a continuing dialogue between the Admiralty, the government and the Treasury, where it was in the Admiralty's interests to overstate the case, the Treasury's to understate and the government's to say nothing that would commit them to one particular course of action.

The ending of the Great War had certainly weakened the Admiralty's claim on a large proportion of government income, by removing any European threat and not clearly identifying any foreign naval threat to the British Empire.[5] The Board of the Admiralty attempted to define imperial sea power at a time when the government was hesitant about defining strategic objectives.

Certainly, in 1919, when the Chancellor of the Exchequer recommended a reduction in the number of capital ships in commission with full crews, down to 15 capital ships and not the 21 that the Admiralty wanted, they played on the fears of a naval race with the United States and Japan, the consequent loss of prestige and trade, and the weakened bonds with the Dominions that this could result in. Both Japan and the United States were building powerful battleships and battle cruisers, all judged to be much more powerful than existing British warships. To avoid being left behind, the Royal Navy felt that it would need to build new ships and, in the absence of any firm government direction, Admiral Beatty, the First Sea Lord, and the Board were convinced that the best way to counter both the US and the Japanese naval threats was to maintain the largest possible battle fleet, based around the battleship. However, when Lord Long, the First Lord of the Admiralty, sought a definitive ruling on government policy regarding naval supremacy over the United States and other naval powers, he was instructed to plan on the basis that it was unlikely that Great Britain would be involved in another war for ten years. And in the same year that Grant wrote his paper and the United States was dismissed as a potential enemy, the Cabinet directed that the maximum amount for the 1920–21 Naval Estimates was to be £60 million, an enormous reduction on Long's proposed £171 million and a much more serious consideration for the Admiralty. It seemed that the Treasury, and not the Cabinet as a whole, was now playing a major role in determining defence policy, dictating how much money was available for the three Services, and thus how Britain could react to situations abroad.

According to the Admiralty's interpretation, the British, as the only Western naval power with a major naval base in the Pacific, Hong Kong, could not allow the Japanese to achieve regional superiority unchecked by alliance or other limitations. In the event of a war with Japan, a large battle fleet would not be sufficient without a base in the Pacific from which to operate and the Director of Plans, Captain Dewar, pointed out to Captain Domvile, the Deputy Chief of Naval Staff, that this raised unpleasant political questions which the Cabinet would need to address. Principally, these were whether the Anglo-Japanese Alliance was to be allowed to lapse or not, the possibility of war with Japan, and the provision of a sufficiently powerful battle fleet and the necessary bases, both along the lines of communication and in the western Pacific region itself, to support the fleet in such a war.[6] This would entail a large amount of money being spent, especially on providing a naval base able to support a battle fleet in the Far East, something that did not currently exist anywhere in the region. *Ad hoc* war responses were to be avoided, and in their place there should be detailed plans for which ships to send, where from and where to; a defended naval base was needed.[7]

The first, tentative, stages in developing a strategy for a Far Eastern war had begun and one of the ways that the Admiralty felt could promote regional security and also reduce the costs to the British government was with the idea of a single, imperial navy, with unity of command on the basis of the experiences of the late war.[8] For their part, the Dominion Prime Ministers felt this to be politically unacceptable and instead they requested that a prominent naval figure be sent to the region to advise them on naval matters. The former commander of the Grand Fleet and First Sea Lord, Admiral Sir John Jellicoe, was chosen by the Imperial War Cabinet in December 1918, as the figure of sufficient standing and prestige.

THE INFLUENCE OF FAR EASTERN STRATEGY

Jellicoe's Mission did not advance the debate at all. As he toured the region he became more and more convinced that the Japanese posed the only real threat. Instead of the Admiralty's 'Imperial Fleet', he advocated the formation of strong Dominion navies and a strong Pacific Fleet. Great Britain would be responsible for providing all eight battleships, six battlecruisers, eight cruisers, 28 destroyers and one depot ship, 22 submarines and one depot ship, two aircraft carriers, eight minesweepers and one repair ship, along with a financial contribution of £14,066,800, some 75 per cent of the total cost, whilst Australia would be responsible for the other two battlecruisers, eight cruisers, ten destroyers and a depot ship, eight submarines and a depot ship, the minelayer, four minesweepers, one aircraft carrier and one repair ship and a financial contribution of £4,024,600, or 20 per cent of the cost. The balance would be made up by New Zealand and Canada. It was an 'Imperial Fleet' by another name, with the British, who would make the largest contributions, more than likely to insist upon operational control, a condition the Dominions would be bound to oppose. The Admiralty had only one course of action open to them; they rejected Jellicoe's report. However, they were equally pessimistic about the alternative, maintaining a battle fleet in the Far East, in peacetime, to match the nine Japanese battleships and eight battlecruisers. In the event of a war, British naval forces would be numerically inferior to the Japanese, incapable of doing anything more than fighting a holding action until the arrival of the British battle squadrons.[9] The solution was for Singapore to be developed as a defended fortress and naval base, able to hold out against assault until relieved by the Main Fleet. The Dominions would assume responsibility for providing fuel for the transit of this Main Fleet to the region, whilst the British provided the ships and the strategy to fight a war against Japan. It was in the light of this shift in strategic focus to the Pacific that the Admiralty's planning began to gather pace.

In the immediate future, the Royal Navy remained stronger than the Imperial Japanese Navy in all classes of vessels, with an overall fleet strength of 33 battleships, eight battlecruisers, 60 light cruisers and 352 destroyers, mainly concentrated in home waters. On paper this was a large enough navy to send a battle fleet eastwards and maintain naval superiority in home waters. But many of these ships were old, had seen arduous war service, and would need refitting or replacing, a process already planned for by the Admiralty, but dependent on high naval expenditure and likely to prompt another new naval race. Strategically it was also felt to be unwise to divide the fleet, and instead it would have to be concentrated into the Main Fleet, a mutually supporting, balanced fleet of all elements, based around the battleships of the Atlantic Fleet and Mediterranean Fleet, detached forces of cruisers securing the sea lines of communication and local defence forces, again, mainly cruisers, on foreign stations.

There were other reasons, apart from the purely strategical, as to why these deployments made sense. Capital ships operating on foreign stations missed out on the opportunity of exercising with other battleships in the same battle squadrons and fleets, with detrimental effects on their own and their battle squadron's gunnery and overall performance; and, more significantly, the dockyard facilities were absent. Even Malta had difficulty in supporting bulged battleships until the arrival of an ex-German floating dock in 1926, and until Singapore was completed, sending any capital ships there was not a realistic deployment.[10]

Virtually all of Britain's battleships were worn out by war service and as battleships remained at the core of the Admiralty's strategic plans, the Board of the Admiralty warned the Cabinet that if Britain failed to complete new capital ships and both the Americans and Japanese completed their programmes, the Royal Navy would become unable to defend Britain's imperial interests. Beatty, the First Sea Lord, wanted arguments to support his position regarding large ships and comparisons with Britain's rivals provided a straightforward and easily identifiable justification for a large Royal Navy battle fleet:

> Unless we start building now, we shall have only the *Hood* in 1925. The United States by 1925 will have 12 Capital Ships of over 40,000 tons of post-Jutland design, embodying war experience in an important degree, as well as 4 such ships of over 30,000 tons. Japan will have at least 8 such ships, most of them over 40,000 tons, by 1925, and by 1928 will have 16.[11]

But, in 1920, visits by W.F. Marriott, the Naval Attaché, to the Japanese imperial dockyards and private shipping establishments seemed to confirm Britain's short-term superiority over the Japanese. Marriott visited Kure, Sasebo, Miadzum, Kobe and Nagasaka, as well as the Etajima Naval College, and came away with the opinion that Japanese naval construction was proceeding slowly because of the expense, labour difficulties, limited resources and lack of raw materials. The dockyards were hard to supply, being in out of the way places, chosen for their seclusion and anchorages, not for their supplies of labour. He wrote in his report that

THE INFLUENCE OF FAR EASTERN STRATEGY

> The keynote of all the Imperial dockyards is lack of funds and the increased cost of labour and materials which is making progress such a very difficult matter. Nowhere during my visits did I notice great activity being displayed. It appeared that all work was going along very leisurely, and I was informed that the dockyard workmen are not giving satisfactory results.[12]

The message he came away with was that the Japanese Navy was in need of money, a proper submarine school, enough wireless stations to maintain adequate communications and a naval air force. In short, not a force to fight a sustained, naval campaign.

This was not the message that Beatty wanted. The message that the Admiralty gave to the politicians, despite Marriott's reports, was clear though: the Royal Navy needed battleships, it needed new battleships and it needed them to be authorized as soon as possible in order to maintain naval superiority. In the short term, Beatty felt that the Royal Navy could cut its budget as the Treasury wished and run on war-surplus stocks until the 1923–24 Estimates. These Estimates would then become crucial, as the ships and equipment of the Fleet would need refitting and replacing if the Royal Navy was to remain superior to other navies.

Preliminary recommendations were for a post-war navy of three fleets in home waters – the Atlantic Fleet, the Home Fleet and the Reserve Fleet – of which only the Atlantic Fleet was to consist of ships in full commission. This would total 11 battleships, five battlecruisers, six cruisers, three destroyer flotillas of 18 destroyers each, five submarine flotillas of six or seven boats each and a flying squadron of the aircraft carrier *Furious* and six seaplane carriers. The Home Fleet would have a further six battleships, six cruisers and three destroyer flotillas, partially manned. Overseas, the Mediterranean Fleet would have a further six battleships and six cruisers, with one flotilla of destroyers and one of submarines, whilst the China Fleet would have one battlecruiser, four light cruisers, one flotilla of destroyers and one of submarines. Other naval stations would have cruisers or minor warships stationed there.

With no European challenge on the horizon, Beatty, along with many others, continued to identify Japan, with its growing naval strength and clear intentions of taking advantage of China's internal weakness, as the most likely threat to British interests. The Admiralty urgently sought a clear statement of whether the Anglo-Japanese Alliance was to be renewed, and what the government's future strategic policy was in order to make contingency plans in the event of war, and to plan the future fleet.[13]

The Admiralty also sought to use government directives such as the 'Ten Year Rule' as a means to plan Britain's long-term naval strategy and the ships to implement it, aiming to build up a balanced battle fleet, on the assumption that the earliest there would be a war would be the early 1930s. Once President Harding had issued the invitations to attend the Washington Conference, the British were forced to rethink both their naval and their Pacific policies, because although they were willing to have the threat of a naval race with the United States removed, and thus more than happy to agree to parity, they wanted to maintain sufficient superiority over the Japanese, so that in the event of war it would be possible to send a fleet to the Far East.

Unsurprisingly, therefore, there were problems initially over agreeing to limit the numbers of battleships each power was to have in commission, and over questions of tonnage and gun calibre. The Imperial Japanese Navy, for example, had long held the view that their battle fleet should be composed of eight modern battleships and eight modern battle cruisers, and the latest version of the 8:8 Plan had 16 fast, modern ships with a Reserve Fleet of two old battleships and three armoured cruisers.[14] In agreeing to the Washington proposals, Japan was effectively agreeing to give up its first line of defence according to the 8:8 Plan, and was looking for some sort of compensation for this. This was the regional superiority enshrined in the Washington Treaty, and neither the Americans nor the British were allowed to develop new naval bases in the region, or improve the defences of any older ones. The exception, which the British negotiated hard to keep out of the Treaty, was Singapore. The Committee of Imperial Defence concluded that no agreement should be reached at the Conference to interfere with the development of Singapore which was overdue, and was claimed to be for defensive, as opposed to offensive, purposes, unlike Hong Kong, being over 3,000 miles from Japan.[15]

The ten-year building holiday on battleship building that was also agreed on was exactly the restraint for which the governments of both the United States and Britain had been hoping. For the next ten years the British would know exactly how many battleships each of the navies would have, maintaining the naval *status quo* and allowing time for the rest of the vessels necessary for a balanced battle fleet, the aircraft carriers, cruisers, destroyers and submarines, to be ordered and launched.

But with the British government more inclined to rely on the concept of collective security and joint action by the League of Nations, and with the lack of any readily identifiable enemy, the Admiralty still felt deeply uncertain as to how to maintain a large battle fleet, other than by emphasizing the threat to Britain's imperial possessions in the east, the vital necessity of maintaining a large and balanced fleet to meet this threat, and the unique position of Britain, as the only naval power other than Japan likely to have a large, first-class naval base in the region.

Unable to build new battleships, the Admiralty set about making the existing battle fleet as potent a weapon as possible through modifications and refits, and between 1918 and 1924 the Royal Navy had the ships of the *R* class bulged and the *Queen Elizabeth* and *Repulse* classes taken in hand for modification, making improvements in machinery, adding bulges or blisters, improving gun elevations and fire control systems and modifying secondary batteries. And just before *Nelson* and *Rodney* were commissioned in 1927, the three *King George V* class battleships were paid off, two for scrapping and one for use as a radio-controlled target ship. As a result of this, the Royal Navy's battle fleet remained at 20 units, all reckoned to be superior to the battleships of the United States Navy and the Imperial Japanese Navy.

During 1922–23, however, the Admiralty refined its interpretation of the One Power Standard and the Ten Year Rule, taking the latter to mean that there was a period of grace in which to develop the base facilities, war plans and fleet for a war with Japan. Sketch Estimates of £61,850,000 were approved by the Board of the Admiralty to maintain the size of the battle fleet, develop the Singapore base and provide up-to-date fuel reserves. And as Bonar Law's government was unwilling to compromise on the concept of imperial maritime security, Admiralty programmes were secured more or less intact. Treasury objections to this were pushed aside, allowing the Admiralty to develop its policies.

With the building of battleships stabilized by Washington's Ten-Year 'Holiday', the new naval rivalry centred on cruisers. The Washington Treaty encouraged the growth of the cruiser by not setting ratios, as had been done for battleships, and encouraged all naval powers to build up to the maximum of 10,000 tons and an 8-inch gun armament.[16] As a consequence, the Admiralty had to work out some sort of formula for the numbers of cruisers it felt were required by the Royal Navy. Cruisers were going to be an important asset in a Far Eastern war, both for fleet work and for trade protection, and initial Admiralty claims calculated that the minimum cruiser force necessary was 69 ships, less than the preferred total of 100 based on recent war experience. However, Balfour had sent a telegram from Washington in November 1921 which gave a different basis for deciding on cruiser requirements:

With regard to cruisers we propose maximum size of individual ships should be 10,000 tons and are prepared to accept five, five, three ratio for number of cruisers for fleet purposes provided our claim is admitted to a substantial surplus for the defence of the Empire trade routes.

Total number for all purposes tentatively suggested by First Sea Lord, fifty cruisers i.e., six less than now, but do not propose to say this at present.[17]

The Naval Staff considered 56 was too low a number and that 70 cruisers were essential for imperial defence duties. Captain Dudley Pound, the Director of Plans, calculated that a total of 17 Class-'A' cruisers, armed with 8-inch guns, and 53 Class-'B' cruisers, armed with 6-inch guns, was the minimum figure acceptable, using the Imperial Japanese Navy, and not necessarily the tasks the Royal Navy had to perform, as a comparison.[18] A serious analysis might have resulted in a smaller fleet, whereas the greater the superiority in cruisers the Royal Navy could achieve in a war with Japan, the more effectively Japanese trade would be halted and the quicker its fleet would be brought to battle in its defence and defeated. The defeat of Japan, it was confidently believed, would follow soon afterwards.

But, according to the Admiralty's figures, the Royal Navy was short of some 22 vessels. In 1919 there had been 84 cruisers afloat, but since then 36 of these ships had been scrapped. Of the remaining 48, ten were over-aged, all were built during or before the Great War and 18 were better suited to North Sea and not Far Eastern operations. But by the time that any new ships reached the fleet, a total of 15 British cruisers would be over-aged and any failure to build replacements would lead to an erosion of Britian's cruiser numbers and of the ability of the Royal Navy to fulfil its roles.

In 1922 the Japanese effectively started a naval race in cruisers when they laid down two units of the *Furutaka* class, with six 7.9-inch guns, capable of tackling a US *Omaha*-class cruiser or a British *Hawkins*-class ship and, with a displacement of 7,100 tons, well within Washington's limits.

The British reply was the *Kent* class and the 1923 Naval Estimates allowed for eight of the new, 10,000-ton, 8-inch gunned cruisers, built to match the new US and Japanese cruisers, as well as three large submarines, two depot ships, a minelayer, two destroyers and an aircraft carrier. And although the Treasury agreed to authorize the funding for this programme, in the first of many delays and cancellations, the new Labour government of Ramsay MacDonald would only authorize five of the cruisers and two destroyers, at the same time delaying the work on the new dockyards at Singapore. It was left to Stanley Baldwin's Conservative government, re-elected in November, to authorize an amended Estimate in 1925 which included all eight cruisers, five submarines, a destroyer, an aircraft carrier and 300 aircraft.[19]

But the Admiralty's assumptions about a war with Japan were challenged by Winston Churchill, the Chancellor of the Exchequer and former First Lord who, along with others in government, was quick to dismiss any fears of a war against Britain by Japan.[20] The Chancellor was looking for economies in the Services' budgets and voiced his doubts about the Japanese constituting a serious threat, as well as challenging the Admiralty about their strategic plans for a war against Japan. He wanted economies, including a programme of fewer ships, more suited to escorting convoys than scouting for the battle fleet and engaging in a fleet action, and in 1925 Churchill stated that if such a war did materialize then the British would be able to sweep the Japanese off the seas in three to four years without the large numbers of cruisers the Admiralty stated were vital.

A committee under Lord Birkenhead examined naval expenditure and recommended cuts; the Admiralty was 'on the ropes' and only the threat of mass resignation by Beatty, Bridgeman, the First Lord, and the entire Board of the Admiralty resolved the matter, allowing the Admiralty to lay down the cruisers they considered necessary for defending imperial sea lines of communication. Baldwin's government approved the construction of seven cruisers for the 1925–26 Building Programme and one light cruiser, eight destroyers and six submarines for the 1927–28 Programme, with the promise of a similar number of destroyers each year until the 1929–30 Building Programme. Construction of the aircraft carrier was deferred until more experience of naval aviation had been gained, using the smaller *Argus*, *Eagle* and *Hermes*, and the larger *Furious*, *Glorious* and *Courageous*, once they were reconstructed and with the battle fleet.

Thus, by 1925, the Royal Navy was still able to send a balanced and potent fighting force to the east, if required. A plan for a war with Japan, War Memorandum (Eastern), had been written the previous year and the defended dockyard at Singapore, a key component of this strategy, was under way, albeit slowly. Without Singapore, any ships needing refit, or damaged in fighting in the Far East, would need to go to Sydney, Simonstown, Bombay or even Malta for repair, involving a considerable sea journey.

There were also two battleships, *Nelson* and *Rodney*, under construction, along with five new cruisers, whilst three large aircraft carriers were also planned for the fleet before the end of the decade. Naval Estimates stood at over £360 million and the Ten Year Rule gave the Admiralty a yardstick for planning, with a date of 1933 as the earliest that war was to be expected. As Chatfield, the Controller of the Navy in 1925, clearly showed, this was another important factor in the Admiralty's planning for a possible war in the Far East:

> This gave the Admiralty a lever. Ships took a long time to build, and dockyards to restore to life. They could lay their plans for the defence of the Empire accordingly, they could resist attempts to cut them to the bone because they could properly argue that such and such a cut would be one that could not be recovered from by 1933.[21]

But the internal battle between the Treasury, with its demands for economy, and the Admiralty, with its war plans, was far from over:

> In the long term, however, the navy's position was substantially weakened by the outcome of this debate over naval policy. Most importantly, Churchill obtained a Cabinet ruling that the Admiralty did not need 'to make preparations for placing at Singapore for a decisive battle in the Pacific, a British battle fleet with cruisers, flotillas and all ancillary vessels superior in strength or at least equal to the sea-going Navy of Japan'.[22]

The Cabinet 'fudged' the issue, accepted that the Japanese threat was a remote one, but that the Admiralty should continue long-term planning, with the proviso that there would be no immediate need to send a fleet eastwards. The Admiralty, for its part, continued to plan on the assumption that war was likely from 1935. Even when, in 1928, Churchill successfully had the Ten Year Rule made self-perpetuating, with no definite date beyond which war was more likely, the Admiralty, undeterred, merely continued war planning against Japan. They just ceased to refer to this officially.

However, further fundamental changes to the strategy also occurred in 1928, when Admiral Beatty retired from the post of First Sea Lord. Vice-Admiral William Fisher, the Deputy Chief of the Naval Staff, signalled a departure from the thinking that had dominated when he suggested that it would make more sense and be more convenient to adopt a more general line of strategic planning. Instead of concentrating on a war with Japan (something the Admiralty had been instructed not to do, and which would stretch the Royal Navy to the limit), more general problems better suited to a smaller battle navy, such as the defence of communications and the mobility of the fleet, should be addressed. Admiral Sir Charles Madden, the new First Sea Lord, agreed, although he did not feel that the time had quite come to make this break. The Admiralty was in fact becoming increasingly adept at defending its policies without referring to the possibility of a war with Japan, and until the onset of Japanese aggression in Manchuria in 1931 and Shanghai in 1932, and certainly until the rise of European dictatorships made planning much more complex, this remained the case.

If 1925 was a high point for the Admiralty's plans, it was also the start of the decline. In the run-up to the 1927 Geneva Conference, for example, two cruisers were cancelled as a gesture of faith by the British government. Cruisers remained, and would remain, the major issue for the Admiralty during the 1920s and most of the 1930s. None of the Washington signatories was happy with their first-generation cruisers. They were big, powerful and expensive. As long as no battleships were being built, building cruisers was helping to keep ailing shipbuilders in business but, once the naval holiday was over, building such ships would create funding problems, and priorities needed recasting in the light of this. So the Admiralty's priority was for numerous, small, modern, fast cruisers, suitable for trade defence and working with the fleet, and some 8-inch gunned ships to match an enemy's cruisers.

Many of the Royal Navy's existing cruisers were old and small, designed for fighting the Germans in the North Sea, not the Japanese in the China Seas and, not unsurprisingly, with the exception of the ships being built, were judged by the Admiralty to be inferior to modern, foreign construction. By 1926, the Admiralty had decided that no cruiser of 10,000 tons displacement could both fight with the fleet and wage trade warfare. Oceanic work required endurance and seaworthiness; combat demanded speed and gun power. On 10,000 tons either demand could only be met at the expense of the other and of armour protection.[23] Nevertheless, the British still needed to have some ship able to match foreign 8-inch cruisers, as the more 8-inch cruisers there were at sea, the more likely a British cruiser was to meet one in single-ship combat, and an older or smaller 6-inch cruiser would be at a distinct disadvantage.[24]

Both the United States and the Japanese, with their eyes on a clash in the Pacific, wanted large cruisers; the United States Navy wanted ships with sufficient endurance to cross the Pacific, whilst the Japanese wanted big cruisers carrying as many guns as possible for the decisive battle they expected to fight.

In an effort to resolve this issue, the naval powers came together at Geneva in 1927 to try to reach agreement. The US delegation proposed a ceiling of 300,000 tons for their navy and the Royal Navy, making a total number of 30 of the 'Washington' cruisers. This was nowhere near the number the British felt they needed and they proposed, in turn, an exemption, because of the demands of providing cruisers for imperial duties. Britain proposed a smaller, 7,500-ton cruiser, allowing 70 for the Royal Navy, 47 for the United States Navy and 21 for the Imperial Japanese Navy, a ratio of 10:7:3. Trade defence was not actually the issue. Attacking trade, Japanese trade, was.

> The Naval Staff knew that Japan could not cripple the trade of any enemy during any war. By acquiring cruisers able to withstand this danger, however, the Royal Navy could gain the means to win a war of attrition against Japan... While the Main Fleet checked the Japanese navy, British cruisers would strangle the Japanese economy, either destroying it, or forcing the enemy to accept a decisive battle.[25]

Cruisers were seen as vitally important to Britain's Far Eastern war plan, and the Admiralty fought hard for numerical supremacy in cruisers. But as neither the United States nor Japan found Britain's proposals acceptable, the conference broke up without agreement.

Another problem was becoming apparent as Britain's overall naval superiority began slowly to decline, as vessels needed refitting, modernization or replacement. In 1928, Parliament was warned by the Admiralty that this was the last year in which costs would be balanced by the sale of obsolete ships and the consumption of war surplus stocks. It was also warned that the Admiralty was presuming that the stable naval situation would continue only up until 1929, the end of the initial ten-year period imposed by the Cabinet in 1919. Once the Ten Year Rule had been made renewable, long-term planning became both more difficult and harder to justify and the outbreak of war was pushed back until 1938, at the earliest.

To many contemporaries, the basis of British naval strategy during this time should have been the maintenance of a concentrated battle fleet and a large number of dispersed cruisers, not one of parity with the United States, but of adequacy for Britain's imperial responsibilities, as defined by government. However, naval conferences and arms limitations had a definite appeal to politicians and the electorate, who were more interested in peace and social reform, leaving the Royal Navy increasingly unable to implement the Admiralty's war strategy. And when, in May 1929, the Conservative Party lost the election to Labour and Ramsay MacDonald became Prime Minister again, one of his first acts was to consult with President Hoover about further reductions in the size and numbers of warships of all types, especially cruisers. In February 1930, the government published a White Paper calling for the curtailment of the current building programme coupled with a reappraisal of British defence needs, and counter-balancing them with a more active role for the League of Nations. Battle fleets, it was hoped, would be kept in a state of equilibrium by frequent naval conferences based on the

1. The traditional view of British sea power; British battleships in line ahead.

2. A Sopwith *Cuckoo* drops a torpedo on the Atlantic Fleet in Portland Harbour, 6 September 1919. Despite scepticism about Scott's views, exercises such as this showed the Royal Navy's early commitment to using aircraft at sea.

3. HMS *Queen Elizabeth*, name ship of her class and for many years the core of the proposed Main Fleet to be sent to the Far East.

4. HMS *Resolution*, of the less well-regarded *R* class, adequate for the Far East, in Pound's view, but described by Admiral Somerville in 1942 as having little fighting value.

5 and 6. Two views of HMS *Courageous*, in company with HMS *Furious* and HMS *Glorious*, working together as a squadron.

7. One of the Royal Navy's *County*-class cruisers. Ultimately judged to be too big and too expensive to build in large numbers for Britain's extensive imperial commitments.

8. HMS *Dauntless* in the 1920s, with a hangar just forward of the bridge. *Dauntless* and the other *C*- and *D*-class cruisers were always unsuitable for the extensive operations judged necessary in the Far East, having being built with North Sea operations in mind.

9. With a heavier armament than HMS *Dauntless*, longer range and better habitability, ships such as HMS *Leander* were much more suited to Britain's need for large numbers of cruisers.

10. The final, pre-war cruisers, such as HMS *Southampton*, were designed with fighting the Japanese 8-inch cruisers in mind. They were expected to be able to defeat their enemy with a higher rate of fire from their main armament of 12 6-inch guns.

11. Seen here with 'a bone between her teeth', ships like HMS *Express*, with four 4.7-inch guns and eight 21-inch torpedo tubes, formed the mainstay of Britain's destroyer flotillas in the 1930s.

12. With their heavier armament of eight 4.7-inch guns and five 21-inch torpedo tubes, the *Tribal*-class destroyers were a response to heavy, foreign destroyers such as the Japanese *Fubuki* class.

success of the Washington Conference, with a total fleet tonnage and limitations on each type of ship. In March the 8-inch gunned cruisers *Surrey* and *Northumberland*, and two 6-inch gunned *Leander*-class cruisers were cancelled, as an act of good faith, in advance of the forthcoming naval conference, and no new construction was to be undertaken until after the conference ended in April.

The results of the 1930 London Naval Conference were mixed. The United States still pressed for more heavy cruisers for operations in the Pacific, whilst the Japanese wanted a 5:5:3 ratio for heavy cruisers and a 10:10:7 ratio for light cruisers. Under political pressure from MacDonald, the First Sea Lord, Admiral Madden, was compelled to accept 50 as a maximum number of cruisers for the British, not the 70–75 cruisers the Admiralty judged necessary. Fighting something of a rearguard action, Madden specified that at least they must be new ships, equal in fighting power to those being built abroad, and habitable in all climates, as well as having a wide radius of action. When it was realized by the government that of these 50 cruisers, six would be Royal Australian Navy and Royal New Zealand Naval Division vessels and 22 would be required for operations with the battle fleet, leaving only 22 for the defence of trade, all it could do was agree.[26]

The scrapping of the *Tiger* and three *Iron Duke* battleships, admittedly old, coal-burning ships of decreasing fighting value (a fourth ship, *Iron Duke*, was retained, with a reduced armament, as a training ship), further weakened the Royal Navy's ability to carry out War Memorandum (Eastern) in the Admiralty's eyes, reducing the total strength of the battle fleet to 15 ships, *Nelson, Rodney, Renown, Repulse, Hood,* five *Queen Elizabeth*-class battleships and five *R*-class battleships, with no margin of superiority over the Japanese if some ships were to be left in home waters. Under the terms of the London Conference, no new construction could start until 1936, by which time many of the British ships would need major refits, or scrapping. Acutely aware of the possibility of sending a battle fleet to Singapore, the Admiralty rejected fresh approaches from the United States in 1932 to reduce the numbers of battleships from 15 to 12, although they were willing to consider limiting new construction to 25,000 tons and 12-inch guns, which would have cost less and released funding to build cruisers. The Admiralty, in fact, prepared designs for such vessels which they could have presented to the London Conference if required, but they were successful in persuading MacDonald that a larger ship could carry more fuel, a heavier armament, do more, be in commission for longer and thus be more economical in the long run.

THE INFLUENCE OF FAR EASTERN STRATEGY

There is little doubt that after the London Conference of 1930 the Royal Navy, with its ageing battle fleet, was being faced with more potential problems. The editors of *Brassey's Naval and Shipping Annual* for 1931 wrote:

> The close of 1930 found the Royal Navy in a condition not altogether satisfactory to those who believe that the Fleet is still the main bulwark of our security. Less money is being spent on it; there are fewer ships in hand; replacements are not being provided for as they become due.[27]

British cruiser strength was 54, an impressive sounding number. But it was made up of *Brisbane* and *Adelaide*, two old cruisers of the Royal Australian Navy, 24 *C* class, eight *D* class, two *E* class and four *Hawkins* class, all essentially obsolete First World War designs. The only modern ships were the 13 *County* class and the two *York* class. The weaker, older vessels would soon start to reach the end of their service lives and in 1931, for example, six would be due for scrapping, but only one new vessel was planned, leaving British cruiser strength at 49. In 1932 six ships were due to be scrapped, with another six in 1933, with only four new ships laid down, leaving a total of 41 British cruisers. If matters were allowed to continue until 1936, by which time 14 more ships would have been disposed of, and only ten completed, British cruiser strength would have declined to 36 – a totally unacceptable number of ships.

The 1932 and 1933 editions of *Brassey's* also highlighted the core problem in constructing cruisers, namely finances, even though the problem had been recognized and spoken about in Parliament. Even Ramsay MacDonald was now wondering if Great Britain had been peacemaking a little too long, whilst Stanley Baldwin, in a speech at the Lord Mayor's banquet in November 1932, wondered if it was not time for Britain to end its unilateral disarmament.[28] But as long as the country was in a financial recession there was little that could be done. Spending on the Naval Estimates continued to fall and the strength of overseas squadrons was reduced as an economy measure. In 1924, for example, eight battleships had made up the Mediterranean Fleet, but in 1932 there were only five.[29]

Similarly, speaking at the opening of Navy Week in Portsmouth on 5 August 1933, Admiral Jellicoe told his listeners that the Royal Navy had been reduced in size far more than any other navy except the German Navy, and that they had been forced to reduce by the Versailles Treaty. Now was the time to stop the decline, he felt, a sentiment echoed in the 1934 edition of *Brassey's*, which deemed both the Washington and the London naval treaties to be crippling British naval power, and called for the problems to be addressed.

This was the situation facing Chatfield when he became First Sea Lord in 1933. As Chairman of the Chiefs of Staff Committee, Chatfield saw the rebuilding of all of the armed forces as a priority, and his colleagues, Field-Marshal Montgomery-Massingberd and Air Chief Marshal Ellington, were equally determined. Their predecessors' report for 1932 had pointed out the unpreparedness of the armed forces for war and their 1933 report was strongly worded, expressing the continued concerns of the Chiefs of Staff.

As a result of this, the Defence Requirements Committee, consisting of the three Chiefs of Staff, Sir Robert Vansittart of the Foreign Office, Sir Warren Fisher of the Treasury and Sir Maurice Hankey, started work in November 1933 and produced a report in February 1934 pointing out the weaknesses in each of the armed services. A second Defence Requirements Committee was set up to report on what was necessary to improve the country's defences. Among these was a new standard of naval strength, taking into consideration the Royal Navy's wide responsibilities. However, the sticking point remained financial.[30]

In Europe, the next Geneva Conference to discuss further limitations was adjourned without agreement in 1934. The rise of the Nazi Party made Germany a factor in strategic planning, demanding, as it did, equal rights and the cancellation of the Versailles Settlement. The development of the German *Panzerschiff*, armed with 11-inch guns on a cruiser's tonnage, had been causing particular concern to naval strategists in Britain and France, and with the Japanese and Italians also expanding their fleets, the Admiralty's war plans had to be revised to reflect the ability of the Royal Navy to meet the threat from Japan whilst maintaining an adequate force to match any Italian or German aggression. From 1933, therefore, Britain, with a fleet effectively big enough to face one enemy, was forced to accept the possibility of being dragged into a three-ocean war with Germany, Italy and Japan.[31] There was still a sense of inertia in government foreign policy, a sense of waiting for developments to unfold in Europe and the Far East before developing a foreign policy. Partly in response to this, and the growing threats to Britain's naval position, in 1934 Chatfield, as First Sea Lord, began to consider developing a new, two-power standard navy able to defend the imperial sea lines of communication, Far Eastern interests and, additionally, European interests.

This was first mooted to the Defence Requirements Committee in 1935, the same year in which the Admiralty had become a supporter of appeasement, through its advocacy of the Anglo-German Naval

Treaty, securing superiority over the planned German Navy, which Chatfield saw as a largely pragmatic way of ending a potential arms race before it had begun.[32] By agreeing to a German fleet of five battleships, 21 cruisers, 64 destroyers and a percentage of Britain's submarine total, the Royal Navy was guaranteed superiority over the German Navy, although, in reaching this agreement, Britain effectively aided Hitler's aim of ending the Versailles Treaty as the brake on German rearmament. From Britain's point of view, though, it did allow for a suitable margin of naval superiority to allow for a fleet to be sent eastwards to Singapore if necessary. The treaty with Germany had, at best, minimized the threat that German naval expansion would limit the size of any battle fleet that could be sent to Singapore in the event of a war with Japan. It had also bought some time in which Britain's new naval construction programmes could be completed.

But, even so, just to maintain a battle fleet of 15 ships, the British would need to lay down seven new ships between 1937 and 1939, followed by another five battleships between 1939 and 1943. With the modernized vessels, this would allow a fleet of 11 battleships to be sent eastwards, with a further nine vessels retained in Home waters.[33]

However, the 1935 Italian action in Abyssinia made the strategic situation worse. Not only was it a direct challenge to the League of Nations' notion of collective security, already showing cracks after Japan's invasion of Manchuria, but it was also an important strategic consideration for the British. With Japan and Germany out of the League, and the United States never in it, it would fall to the British, and mainly to the Mediterranean Fleet, to impose any League sanctions against Italy, and the British did not feel strong enough in the Mediterranean to carry this out. Malta and Alexandria, and the Mediterranean Fleet itself, were all vulnerable to air attack, and the Mediterranean sea route to India and the Far East was also under threat. Any action in the Mediterranean would inevitably result in loss or damage of ships, thus limiting the Royal Navy's ability to send the maximum strength eastwards if a crisis with Japan escalated into war. Clearly, the shortcomings of Britain's strategic policy were becoming more and more noticeable.

One more attempt at naval arms limitation took place in London in 1936. The Foreign Office was anxious to placate the United States over cruisers, and suggested a new maximum number of 60 to be aimed for by 1942 for the Royal Navy, whilst the Admiralty wanted to increase the numbers of cruisers back to 70, a case over which Chatfield vigorously argued with MacDonald. There was also the question of the capital ships, long overdue for replacement or modernization, which would swallow up some £120 million. But this

London Conference was not a success, principally because Japan's delegation walked out in January 1936, effectively ending the era of naval arms limitation. The remaining delegates agreed that if Japan refused to sign the agreements by the end of 1936 then the main armament of battleships could be raised from 14-inch to 16-inch. Other minor adjustments were: cruisers were not to be built over 8,000 tons, submarines were to be limited to 2,000 tons, and unrestricted submarine warfare was banned. However, by this time the Royal Navy was in dire need of new battleships (Chatfield had been calling for a new class of modern battleship since 1933, and in his speech at the Cutlers' Feast in Sheffield[34]) and, rather than wait for a redesigned gun, they went ahead with the 14-inch design, and by 1937 the first two of the new battleships, *King George V* and *Prince of Wales*, had been ordered, along with a massive naval rearmament programme, which eventually included all five *King George Vs*, five aircraft carriers and 23 cruisers of various classes.

Even this fleet might not have proved itself large enough to fight a war in two hemispheres. The Defence Requirements Committee estimated costs of some £88 million to £104 million in 1937, which would only fund construction of a fleet of 15 capital ships, eight aircraft carriers, 70 cruisers, 16 flotillas of destroyers, 55 flotillas of submarines and attendant auxiliary vessels, and not the optimum, 'New Standard' fleet of 20 capital ships, 15 aircraft carriers, 100 cruisers (including ten belonging to the Dominions), 22 flotillas of destroyers (including three Dominion flotillas) and 82 flotillas of submarines, plus their attendant auxiliary vessels, sought by the Admiralty. The Treasury was prepared to allocate only £88 million in the face of competing demands from the other Services to rearm, although when the Munich Crisis had passed but war with Germany was seen as increasingly likely, the Treasury was more receptive to the Admiralty requests for money.[35] Even so, the Royal Navy's strategists had to accept that they could no longer honour their commitments to the Australians and send a fleet to the Far East if the situation in Europe was unstable, which it was appearing increasingly likely to be.

How effective, if at all, therefore, had the Admiralty been in their strategic planning during the period between the two world wars? The single biggest problem throughout the 20 years between the ending of one war in 1918 and the outbreak of the Second World War in 1939 had been financial and strategic. Throughout the period there was little, if any, government direction regarding strategic naval planning. Instead, there was a long-held, understandable desire for naval arms limitation as a way of deciding what form the navies should take, and thus what tasks they could undertake. The

Admiralty had accepted the Washington Treaty as it gave parity with the United States Navy at little cost and superiority against Japan, France and Italy. The moratorium of the Washington Treaty had, at least, removed some of the pressure on the Admiralty, allowing for the development of cruisers and destroyers. As there was no treaty limitations on numbers of cruisers, they became the focus for major new warship construction, and Britain's naval leadership, in turn, focused their hopes upon the construction of cruisers of the largest type and in the greater numbers allowed by the Washington accords.[36]

These aspirations were not shared by government and, as a result of budget cuts, the cruiser programme was cut by a quarter between 1927 and 1929.[37] By 1933 British naval spending was well below the record low set in 1924 – only 6 per cent of government expenditure – whilst in the same period the United States was reportedly spending five times as much and Japan three times as much on their navies.

Nevertheless, with no new battleship construction permitted until 1931 at the earliest, the Admiralty put more emphasis into its 1925 six-year programme, to build cruisers, aircraft carriers, submarines, destroyers and auxiliaries, only proposing to build seven new battleships in 1931, after the end of the original Ten Year Holiday proposed at Washington. This meant that

> By 1929 the navy laid down all the programme considered necessary in 1923–24 with the exception of an aircraft carrier. It had the largest and most modern fleet of cruisers at sea, virtually the only operational carriers in the world, alongside the best squadrons of capital ships.[38]

The Admiralty had also completed 60 per cent of the fuel oil reserve programme and, within three years, Singapore would be completed. A detailed war plan also existed. By skilfully stating its case, and lobbying successive governments, the Admiralty had managed to maintain what was probably the strongest navy in the world, numbering 20 capital ships, as compared to the American 18 and the Japanese ten, giving the British the ability to send 15 battleships to the Far East if a war broke out.[39] However, the economic strains undergone by Britain in 1931, with the consequent cutbacks in defence spending, led to the Royal Navy's becoming less of a potent force in the 1930s.

But in 1930, Franco-Italian antagonism proved irresolvable and both powers refused to limit their fleets. The resultant naval competition between them threatened the Royal Navy's ability to secure European and Mediterranean waters whilst defending Far Eastern interests and, at the same time, it was also clear that naval spending would have to be carefully aligned to the ability of the country to afford it:

> Technically the Admiralty could curb the price of sea power by keeping the overall size of the fleet to a minimum; by decreasing the dimensions of warships which in turn made them serviceable in existing dockyards; and by extending operational lifespans to put off expensive replacement building. Since 1927, for instance, the Naval Staff had proposed to reduce the treaty displacement and main armament of capital ships from 35,000 tons with 16″ guns to 25,000 tons with 12″ guns and to stretch their treaty life spans from 20 to 26 years.[40]

Of the First Sea Lords after Beatty, no one – with the exception of Chatfield, probably the best of the inter-war leaders – had matched his drive and determination to ensure the Royal Navy was ready for war. His arrival at the Admiralty marked the end of the years of decline for the Royal Navy, and, in 1935, with the start of Chatfield's lobbying of the Defence Réquirements Committee for new construction, the battle fleet began to be enlarged and modernized.

Japanese aggression in China caught the British off guard and unable to act decisively; the government's reaction was that nothing was to be done that would precipitate a Far Eastern war. However, Singapore was judged to be secure enough to operate a fleet in 1932 and in 1934 the first Singapore naval conference since 1925 was held to discuss the operational use of the battle fleet in time of war. However, by 1935, a changing European situation, brought to a head by the Abyssinian Crisis and the possibility of a war with Italy, meant that the whole Singapore strategy needed reassessing, and in mid-1936 the Admiralty told the Cabinet that it would probably not be able to defend Singapore if at war in Europe. And in 1937 the Chiefs of Staff assured the Imperial Conference of the continuance of the commitment to defend the Far East, at the same time pointing out the potential danger of Italy in the Mediterranean Sea, lying astride the main sea route to the Far East through the Suez Canal.

Rearmament was finally under way and the older battleships were being modernized, even though this could only proceed slowly, as a sufficiently large battle fleet needed to be available in case of crisis. A treaty also existed with Germany, restricting the size of its navy, qualitative limitations remained with the United States and France, and the new ships of the *King George V* and *Lion* classes were authorized and being built.

So, although modernization was overtaken by subsequent events, the Royal Navy entered the Second World War as still the world's largest navy. Between 1929 and 1935, in the face of growing international tensions, the Royal Navy laid down one aircraft carrier, four cruisers, nine destroyers, six sloops and three submarines. Between 1935 and 1939, new building would gather pace in Britain; two new *King George V*-class battleships were laid down, with three more following in 1937. Two aircraft carriers and seven cruisers were also laid down in each year, along with a total of 34 destroyers, 12 sloops and 15 submarines. The 1938 estimates included another aircraft carrier, seven cruisers and three cruiser-minelayers, whilst the 1939 estimates added another aircraft carrier, two cruisers, another minelayer, 16 destroyers, 20 *Hunt*-class escort destroyers, 56 *Flower*-class corvettes and two sloops.[41]

The Royal Navy also entered the war self-confident in its ability to defend Britain's maritime communications, despite an awareness that there were still problems. This self-belief stemmed from a conviction that British sea power remained a potent weapon which could produce decisive results. Certainly, it is fair to state that the Royal Navy was ill prepared in anti-aircraft and anti-submarine tactics at the start of the Second World War (but so were all of the other combatants' navies), and that existing naval aircraft (with the exception of the Fairey *Swordfish*) would soon prove themselves unsuited to modern warfare. But there were reasons for the low priority afforded anti-submarine and anti-aircraft warfare during the inter-war years that seemed rational and reasonable at the time: ASDIC and the Vickers, Mark M, multiple 'pom-pom' gun. Neither were thought to be perfect, but both were considered to be at least equal to, if not superior to, enemy equipment.

Similarly, the lessons from the mistakes made in the amphibious assault at Gallipoli, which were repeated in Norway, could have been better learnt but, again, the Royal Navy had not envisaged conducting large-scale, opposed amphibious landings and only started planning exercises for beach landings in the late 1930s.

The emphasis had been on a naval war against trade and a decisive fleet battle, but far from being obsessed with a surface battle, another Jutland, there was a virtual revolution in surface warfare, with new tactics and more aggressive Battle Instructions. The study of Jutland exposed previous doctrine to critical analysis and was the impetus behind the enquiries as to whether the battleship was an obsolete warship type and, later, how a future fleet would fight:

> The primary naval challenge for which the Admiralty prepared was an epic clash with the Japanese in the South China Sea. This focused

Royal Navy energy firmly on the improvement of surface warfare capabilities applicable in these clearly anticipated conflicts, and the results were outstanding...[42]

War experience would subsequently show up both the strengths and weaknesses of this pre-war planning.

NOTES

1. T. Saxon, 'Anglo-Japanese Naval Cooperation, 1914–1918', *Naval War College Review*, Winter 2000, pp. 2–7.
2. Ibid., pp. 10–11; D. Geddes, 'The Mandate for Yap', *History Today*, Vol. 43, December 1993, pp. 32–7.
3. PRO, ADM 1/8570/287, 'British Imperial Naval Bases in the Pacific, 21 October 1919', p. 22.
4. C. Bell, *The Royal Navy, Seapower and Strategy Between the Wars* (Macmillan, London, 2000), pp. xvii, 7–8.
5. J. Sumida, 'British Naval Procurement and Technological Change, 1919–1939', in P.P. O'Brien (ed.), *Technology and Naval Combat in the Twentieth Century and Beyond* (Frank Cass, London, 2001), pp. 130–1.
6. PRO, ADM 1/8570/287, p. 22.
7. Ibid., pp. 28–9.
8. Tracy (ed.), *The Collective Naval Defence of the Empire, 1900–1940*; PRO, CAB 23/41, 'Minutes of Imperial War Cabinet, 27 June 1918', pp. 230–2.
9. Tracy, *Collective Naval Defence*, pp. 245–6.
10. J. Moretz, *The Royal Navy and the Capital Ship in the Interwar Period: An Operational Perspective* (Frank Cass, London, 2002), p. 185.
11. PRO, CAB 16/37/2, 'Sub Committee on the Question of the Capital Ship in the Navy, December 1920/January 1921, Vol. II', pp. 185–6.
12. PRO, FO 371/5365, 'Naval Attaché's Visits to Imperial Japanese Dockyards and Private Shipping Establishments in Japan, 12 July 1920'.
13. PRO, ADM 1/8570/287; Roskill's work, *Naval Policy between the Wars, Vol. 1: The Period of Anglo-American Antagonism 1919–1929* (Collins, London, 1968), p. 215, and Bell, *The Royal Navy*, p. 21, also provide examples.
14. M. Peattie, 'Japanese Naval construction, 1919–41', in O'Brien (ed.), *Technology and Naval Combat in the Twentieth Century and Beyond*, pp. 93–5.
15. Moretz, *The Royal Navy*, p. 39.
16. B. Ireland, *Cruisers* (BCA, London, 1981), p. 82.
17. PRO, ADM 116/3445, 'The Washington Conference on Limitation of Armaments and Far Eastern Policy; Cypher Telegram no. 73 from Balfour, 28 November 1921'.
18. *The Roskill Papers*, ROSK 7/34A, 'Cruisers-Policy and Construction: ADM 1/8653, Naval Building Programme – Cruiser Requirements; memo to the DCNS and Controller of the Navy from the Director of Plans, 31 October 1923'; ADM 1/8672, 'Cruiser Replacement Programme, 31 August 1924'.
19. D. Van der Vat, *Standard of Power: The Royal Navy in the Twentieth Century* (Hutchinson, London, 2000), p. 152.
20. I. Gow, 'Anglo-Japanese Naval Relations Prior to 1931' (London School of Economics, International Studies, 1985), p. 22.

21. Admiral of the Fleet Lord Chatfield, *It Might Happen Again: Vol. II of the Autobiography of Admiral of the Fleet Lord Chatfield* (Heinemann, London, 1947), p. 11.
22. Bell, *The Royal Navy*, p. 22.
23. J. Ferris, 'The Last Decade of British Maritime Supremacy, 1919–1929', in Neilson and Kennedy (eds), *Far Flung Lines*, pp. 137–8.
24. H. Bywater, *A Searchlight on the Navy* (Constable, London, 1934), pp. 174–5.
25. Ferris, 'The Last Decade', pp. 138–9.
26. R. Higham, *Armed Forces in Peacetime: Britain 1918–1940: A Case Study* (G.T. Foulis, London, 1962), pp. 136–7.
27. Commander C. Robinson, RN, and H.M. Ross (eds), *Brassey's Naval and Shipping Annual, 1931* (W.L. Clowes, London, 1931).
28. Commander C. Robinson, RN, and H.M. Ross (eds), *Brassey's Naval and Shipping Annual, 1932* (W.L. Clowes, London, 1932).
29. Commander C. Robinson, RN, and H.M. Ross (eds), *Brassey's Naval and Shipping Annual, 1933* (W.L. Clowes, London, 1933), p. 12.
30. Chatfield, *It Might Happen Again*, pp. 77–84; Higham, *Armed Forces in Peacetime*, gives the construction costs of a *King George V*-class battleship as £10,000,000.
31. Moretz, *The Royal Navy*, p. 85, referring to Richmond's observations on the illusion that the British Empire, 'a Two Hemisphere Empire can be defended by a one hemisphere Navy'.
32. Chatfield, *It Might Happen Again*, pp. 73–6; M. Simpson (ed.), *The Somerville Papers: Selections from the Private and Official Correspondence of Admiral of the Fleet Sir James Somerville, GCB, GBE, DSO* (Navy Records Society/Scolar Press, Aldershot, 1995), p. 102.
33. Bell, *The Royal Navy*, p. 28.
34. Chatfield, *It Might Happen Again*, pp. 66–9, 72, 121.
35. Bell, *The Royal Navy*, pp. 33–4.
36. Sumida, 'British Naval Procurement', in O'Brien (ed.), *Technology and Naval Combat*, pp. 130–1.
37. Ibid.
38. Ferris, 'The Last Decade of British Maritime Supremacy 1919–29', in Neilson and Kennedy (eds), *Far Flung Lines* p. 140.
39. Ibid.
40. J. Maiolo, *The Royal Navy and Nazi Germany, 1919–1939: A Study in Appeasement and the Origins of the Second World War* (Macmillan, London, 1998), p. 14.
41. Van der Vat, *Standard of Power*, p. 171.
42. D. MacGregor, 'The Use, Misuse and Non-use of History: The Royal Navy and the Operational Lessons of the First World War', *Journal of Military History*, Vol. 56, No. 4, 1992, p. 603.

2

A Far Eastern Strategy: War Memorandum (Eastern)

> Strategy may be divided into two parts, war strategy and preparation strategy; and of these two, preparation strategy is by far the more important.
>
> War strategy deals with the laying out of plans of campaign after war has begun, and the handling of forces until they come into contact with the enemy, when tactics takes those forces in its charge. It deals with actual situations, arranges for the provisioning, fuelling and moving of actual forces, contests the field against an actual enemy, the size and power of which are fairly well known...
>
> Preparation strategy deals with the laying out of plans for suppositious forces against suppositious enemies... war strategy is merely the child of suppositious strategy.[1]

Strategy, the art of employing military force to achieve the aims of national policy, must invariably be planned in peacetime, to provide the framework for action in time of war and for the provision of the relevant armed forces. The development of the Royal Navy's strategy for a war with Japan – War Memorandum (Eastern) – did just this, allowing the Admiralty to determine their most likely enemy and the fleet that would be needed to defeat them in battle. This was the 'preparation strategy', and the Japanese were the 'suppositious enemies'. The 'suppositious forces' were, in this case, the ships that the Admiralty wanted to enable them to carry out this strategy, the naval base at Singapore and, most importantly, the logistical support for such an action. From 1919 the Plans Division of the Admiralty had routinely assessed the situation in the western Pacific, producing war plans to counter a supposed Japanese threat, and attempted to anticipate what could be done to ensure that the whole plan did not break down under the stresses of war.

Numerical strength in comparison with the Japanese was also assessed, once battleship ratios had been settled by the Washington Treaty, a factor that actually worked in the British strategists' favour. But, at the same time, they were conscious that the size of the opposing fleets would not be the size of available ships in the opposing fleets, but the numbers that could be deployed and supported. The further away from home waters the British operated, the more pressing would be the need for refit and repair, and hence the need for a major naval base in the region.

As the Deputy Chief of the Naval Staff (DCNS), in 1919, Captain Domvile had pointed to the possibility of a war with Japan, and the fact that any strategic considerations raised political questions about Japan's future as an ally and that the naval planners needed to know what was going to happen with the Anglo-Japanese Alliance. If it was allowed to lapse, war became more of a probability, and strategic decisions had to be made in an imperial context to allow strategy to be developed before a crisis arose.

The 1924 War Memorandum (Eastern), which developed from the Plans Division's various memoranda, embodied the Admiralty's own assessment of the risk in the Far East, from a naval perspective, the policy necessary to counter this and the fleet needed to implement the plan.[2] Indirectly, it also had a bearing on the tactics the battle fleet would employ and there were several Far Eastern-related assessments made from 1919. Captain Dewar, Director of Plans, pointed to the difficulties of waging a war with Japan in the region, the principal one being that the Japanese would have all the advantages of being unopposed until the arrival of the Main Fleet, estimated in 1919 to be as long as 90 days after a declaration of war, during which time the protection of imperial trade by existing forces, principally the cruisers of the China Squadron and the Royal Australian Navy, would become paramount.[3]

Beatty, as First Sea Lord, made sure that the Admiralty's case was regularly aired. Giving evidence to the Bonar Law Enquiry, in January 1921, Rear-Admiral Osmond de B Brock recognized the importance of holding both Singapore and Hong Kong, whilst Rear-Admiral de Bartolomé believed a war against Japan would involve mainly protecting British trade and cutting Japanese trade. To do this with a battle fleet would require a base from which ships could operate to contain the enemy fleet.[4] In the eleventh meeting, held in February 1921, considering the draft report, Beatty emphasized that all plans for a war against Japan were naval and that any war with Japan could not involve the invasion of Japan, only the cutting of its communications with the outside world.[5] At an early stage, therefore, planning was seen as an exclusively naval affair, not one involving either the army or the RAF. They were just expected to provide the requisite forces the Admiralty planners required, when and where they required them.

One of the navy's more outspoken senior officers, Herbert Richmond, in the same session expressed his belief that Japan, even by 1928, and even with its present construction programme, would not be able to mount an invasion against Australia, a common fear, whilst in an earlier session he had expressed the viewpoint that it would also be difficult to land troops on Singapore Island, as coast defence guns and aircraft would make the approach of transports impossible.[6] Instead, Richmond believed that the Royal Navy had to concentrate on defending shipping in the Indian Ocean from raids by Japanese battle cruisers, something that Richmond believed to be much more likely, and much more dangerous.

If a British battle fleet was going to operate in Far Eastern waters, then the security of Singapore certainly needed to be reconsidered. Existing defences, consisting of batteries of 9.2-inch, 6-inch and 4.7-inch guns, were designed to deter bombardments by cruisers and raids of up to 2,000 men, but if the British government was going to use the next seven years, at least, to prepare the fleet, the bases and the oil reserves, there would need to be an improved level of defences. One of the debatable issues around the whole development of the Singapore base would be whether 15-inch guns or aircraft would be a better deterrent to Japanese battleships. Whilst the Admiralty favoured the proven effectiveness of a 15-inch gun battery, Air Chief Marshal Trenchard of the RAF, eager to carve a place in defence strategy for his new force, argued that aircraft were cheaper, more mobile and had a longer range than guns. He advocated a force of two torpedo bomber squadrons, one fighter squadron and a flight of reconnaissance seaplanes as a far more effective defence for the new naval base, able to strike any invasion force up to 150 miles away. In reality, the question was not a case of whether it was to be either guns or aircraft. Singapore needed both. But in the 1920s the debates about whether aircraft truly were able to overcome warships were still continuing, and the Admiralty elected to stay with the proven defence system, the 15-inch gun.

However, it was still uncertain whether Singapore could be defended successfully. A survey of the Johore Straits, carried out in 1921, indicated that landings would be possible almost anywhere in good weather and in many places in bad weather,[7] and the War Office estimated that the Japanese would be able to land some 10–12 divisions on the mainland, north of Singapore Island, requiring a consequent increase of the garrison there to 10–12 divisions, way beyond the army's capability to provide, owing to the many other commitments it had to fulfil. However, Beatty was confident that the Japanese could not land these forces before the arrival of the British battle fleet, now estimated to be able to arrive in around 40 days, and that,

> in the most favourable of circumstances, the Japanese expeditionary force would have the maximum of 23 days after the landing in the Malay Peninsula to fulfil its mission. Even if the British intelligence system were not reorganized, the maximum time available for the Japanese would be only 34 days.[8]

Unable to afford a large garrison, the War Office was happy to be convinced by the Admiralty that the Japanese would be unlikely to be able to land an invasion force on Singapore Island and capture the base before the arrival of the Main Fleet, a hypothesis which would apparently be confirmed by the 1924 exercise on Salsette Island (see Chapter 3).

Logistical considerations were another issue to be considered, as a British battle fleet had never operated at such a distance from home waters since the days of sail. The original plans were only for deploying warships; the battle fleet would sail east, meet the Japanese battle fleet and defeat it in a decisive battle. But as the logistical problems became more apparent, time had to be spent in building up resources and the strategy had to change to reflect this need for a slower build-up of overwhelming forces in the western Pacific, so that

> by the mid 1920s its horizons had widened to include associated strategic matters such as detailed timetables and tactical information ... Emphasis was laid on the despatch of a fleet to Far Eastern waters, the overall objective being the protection of all Britain's eastern possessions and the decisive defeat of Japan.[9]

Importantly, in May 1921, Brock, now DCNS, had pointed out that as the One Power Standard meant it was impossible to divide Britain's battle fleet to meet any potential threat and as mobility was a vital part of naval strategy, adequate fuel stocks on probable lines of communication were becoming important. Oil fuel reserves in the United Kingdom, amounting to approximately 12 months' stockpiled supply, were adequate but the situation in the Far East and western Pacific was not satisfactory and, until a reserve was in place, it would be impossible to send a fleet to the Far East at short notice, or maintain it on a war footing once it had arrived.[10] So, in an

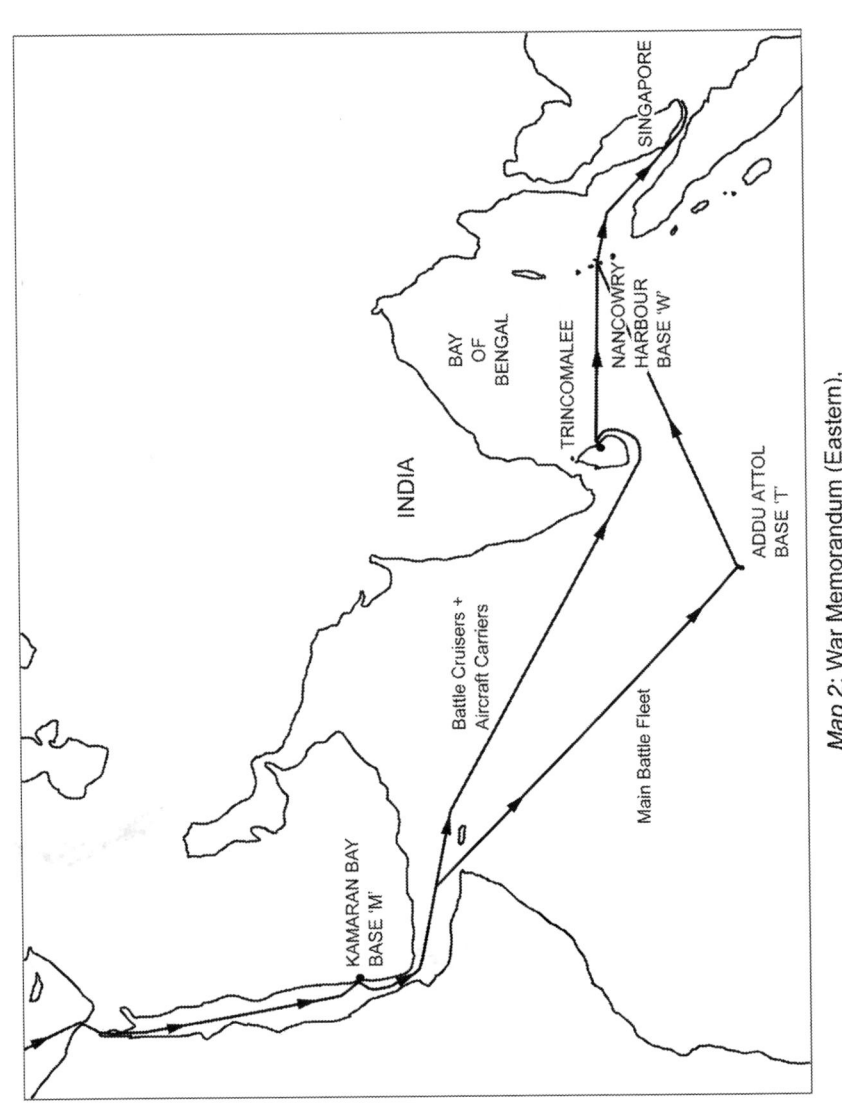

Map 2: War Memorandum (Eastern), The Route to the East

assessment, based on assumptions on the probable strength of a fleet to be sent eastwards as being 20 battleships and battle cruisers, 20 cruisers, 80 destroyers, 30 submarines and ten auxiliaries, including aircraft carriers, Brock minuted:

> The twelve months' war consumption of this fleet with 58 attendant oilers is computed to be 3,430,000 tons... It will be seen that for supply to an advanced base, which may be upward of 2,500 miles from Singapore, a further 48 oilers using 460,000 tons would be necessary.[11]

Enormous amounts of fuel would be needed to allow the battle fleet to operate, and the Admiralty recognized that it was this, and not the actions of the Japanese, that was likely to limit operations. In order to ensure the speedy despatch of the fleet, extra refuelling facilities had to be developed in the southern Red Sea and at Colombo in Ceylon (modern Sri Lanka). If, as was expected, there would be three days of growing tension before hostilities broke out, British politicians would be alerted and would start mobilizing the fleet, which would be fully fuelled, armed and provisioned when war broke out and could immediately leave Malta and Gibraltar, at a steady 16 knots, for Suez and the Red Sea. With no delays other than for refuelling *en route*, total passage time was estimated as being no more than 40 days.[12] But with so many variables it was probable that any actual war would start in highly unfavourable conditions for the British, making the existence of a strong, defended base even more vital.

The dominance of Singapore in any strategy was clearly shown, with over 1,200,000 tons of oil storage planned.[13] Adequate reserve stocks of fuel oil were vital, especially if Singapore would not be able to re-supply from the Netherlands East Indies.[14] Owing to the inevitable lack of transport facilities and the waste entailed when large quantities have to be taken distances such as 10,000 miles, it was not considered practicable to utilize the United Kingdom reserve of 4.5 million tons, and reserves – a further 31.5 million tons – would have to be laid down at ports in the Indian Ocean, Singapore and Australasian waters, with a further million for the merchant fleet, stored in tanks, as the Royal Navy would need every available tanker to carry fuel eastwards.[15]

Although the problems of moving and maintaining a fleet in the Far East were great, they were not regarded as insurmountable, and the Committee of Imperial Defence backed the need for a base at Singapore:

> the fact that a large British fleet cannot be maintained in time of peace in the Pacific must not be interpreted to mean that there is no necessity

to develop one or more of the British naval bases situated there so as to ensure the docking and repairing facilities available in those waters keeping pace with the advances made in ship construction; and that unless this is done it will not, in an emergency, be possible for the British fleet to operate in the Pacific.[16]

Only the Treasury representative to the sub-committee, E.W.H. Millar, regarded the danger of Japanese attack remote, and thus not justifying heavy expenditure. He felt Japan would not risk a conflict with the British Empire unless it could count on US neutrality, and he also believed that attacks on China, as well as Hong Kong, would have to be made before an assault on Singapore could be attempted. Millar stated that defence planning should be based on practical, not hypothetical, scenarios and the subsequent cuts in defence expenditure compelled the Admiralty to accept £16 million in the Navy Vote for 1922–23, cutting back on the construction of defences at Singapore, but not stopping the planning.[17]

By comparison, the idea of being well prepared for a war against Japan was one that Admiral Beatty, as First Sea Lord, advocated as a part of his campaign for the largest possible battle fleet. In his address to the 1921 Imperial Conference he had stressed that command of the seas meant control of communications which, in turn, meant the destruction of an enemy's fleet, the real objective of the Royal Navy.[18] In what was a robust outline of his doctrine, he stated that

> whilst the enemy has a force in being it can never be completely contained and the attempt to do so is fraught with the greatest difficulty, which involves great efforts and great expenditure. The command of the sea is determined by the result of great battles at sea. To any naval power, the destruction of the Fleet of the enemy must always be the great object aimed at.[19]

No fleet, British or Japanese, could act in this way without the fuel reserves and a base from which to operate, and the Admiralty maintained that it could only fulfil its wider security responsibilities and have an advantage over the Japanese if it had both these and the ships. The ships of the battle fleet would be the most obvious indication of naval strength both to some naval professionals and to outside observers, including the politicians, but the ability of the Royal Navy to implement even a rudimentary war plan would be the real measure of its ability to defend British interests.

With this in mind, the Plans Division returned to the nature of the problems associated with fighting a war against Japan when the Director wrote about a possible Japanese war:

Table 1
Oil fuel requirements and proposed supply arrangements in the East, 1921

Class of Ship	Number	Estimated Annual Consumption (tons)	Source or Base	Quantity (tons)	Equivalent at Singapore (tons)	Requiring Oilers
Battleships	12	720,000	Singapore	1,200,000	1,200,000	0
Battle cruisers	8	480,000	Rangoon	400,000	468,000	3
Light cruisers	26	624,000	Burma	100,000	–	–
Flotilla Leaders and TBDs	80	960,000	Persia	1,530,000	1,224,000	30
Submarines	30	36,000	Aden	100,000	123,000	2
Various attendant oilers	58	370,000	Hong Kong and Australia	15,000 400,000	15,000 400,000	
Add for repairs						7
TOTAL	224	3,430,000		3,970,000	3,434,000	42

Source: Tracy, *Collective Naval Defence*, p. 294.

> Should we ever find ourselves at war with Japan it would be the first time in the history of the British Navy that the Main Fleet would be permanently based thousands of miles from Home Waters. In this eventuality we should be faced by conditions entirely different from those under which we fought in the late war. As it is the accepted policy that we must be prepared for a war with Japan, it is necessary to try and ascertain how far existing arrangements would fail to meet the requirements of a fleet working under such novel conditions, and to endeavour to reconstruct our war preparations accordingly.[20]

Again, he stressed that any British fleet sailing eastwards would need the support of a string of bases along its route to enable the fleet to reach Singapore and would face real difficulties operating in the region if Singapore was not a fully functioning naval base.[21] The despatch of a fleet larger than the Japanese, ready to move eastwards to Singapore and able to be replenished *en route*, became the Royal Navy's main strategic consideration as the Plans Division of the Naval Staff made their assessments of a possible war with Japan. By comparison, the lack of Japanese bases between Japan and Singapore would compel Japan to place both its battle fleet, and its mercantile trade, at risk, if it moved against Malaya and Singapore.

The Plans Division anticipated that Japanese trade would soon be stopped in all oceans except the Pacific on the outbreak of war, as Japanese vessels and neutrals carrying contraband would be intercepted by cruiser patrols, and that as soon as the Japanese battle fleet was defeated, trade in the Pacific would also cease. The Dominions' assistance was expected to help achieve this. This was to be the perfect example of the Admiralty's ideal of one, indivisible ocean; an imperial undertaking, controlled from London. Australian and Canadian cruisers and ocean-going submarines would attack Japanese trade, and fuel reserves would help supply convoys and warships. Dockyards in India and Australia would also assist in maintaining and repairing the fleet. Japanese attacks on British trade were expected in the Pacific, but not in other oceans, as Japan had no foreign bases, and its raiders would be hunted down or forced to seek internment. Again, the assistance of the Dominions in providing bases, local convoy escorts, anti-submarine patrols and regulation of shipping near the focal points of their ports was expected.[22]

This was still not a strategic war plan, merely a rudimentary outline of intent, a plan for the deployment of warships which relied on the dispatch of a large fleet eastwards, whilst leaving behind sufficient vessels to deal with any European threat. But it was a start to the planning that would lead to the 1924 War Memorandum (Eastern).

Naval strategy was therefore governed and limited by fuel, and preparation strategy was the way of overcoming this problem. A fleet sent to the Far East would be dependent on its lines of communication for its supplies, and the failure of these would have serious consequences. As long as these lines of communication were secure, superiority over an enemy fleet could be maintained. But ensuring the supplies to the main base and the fleet would make tremendous demands on the reserves of naval stores, and on the ships themselves, both those carrying the stores, and those convoying them. Like war strategy, supply of the fleet must be planned for in peace as it would be too late to improvise later on.

This meant that the planned base at Singapore would have to fulfil all of the functions of Scapa Flow, Invergordon and Rosyth, the major bases in the Great War. There would need to be a protected anchorage and a dockyard, along with repair ships, floating docks, oilers, ammunition and store ships, minesweeping trawlers and patrol craft to keep the approaches secure. There should be sufficient oil reserves at Port Said, Suez, Aden and Colombo to enable the fleet to reach Singapore rapidly from the Mediterranean, and reserves in the Far East, distributed between Singapore, Rangoon and Ceylon, sufficient to allow for several months' operations.

Any fleet operating in the Far East would be some 8,000 miles from Britain, compounding the difficulties of keeping it at a high level of repair and efficiency. Consumption of fuel oil would be some 58,000 tons a week, and if the fleet was at sea for 15 days a month, it would require 200,000 tons of supplies a month, or 2.5 million tons in a year. The total requirements of such a base would be 13–22 shiploads of supplies a week, and up to 60 tankers would be needed to replenish the base with fuel oil. As the sea lines of communication would be at risk from submarine, raider and aircraft attack, convoys would be necessary. Approximately 10,000 tons of supplies other than oil a week would be needed, based on the experiences of the Grand Fleet at Scapa Flow. For every 40 per cent of supplies that went to the fleet, 60 per cent were needed to maintain the base. Also included in the estimates were the provision of guns and ammunition, mines, booms, nets and searchlights for the static defences, aircraft and their spare parts and fuel, anti-submarine and patrol craft, minelayers, minesweepers, stores for the fleet, repair shops, fuel supplies and medical provisions.[23]

As Hong Kong was neither suitable enough as a main fleet base nor far enough north to act as the advanced base for operations, within 600 miles of Japan, an advanced base would be needed, connected to Singapore by cable, and with its own tugs and mooring craft, salvage tugs, vessels capable of lifting heavy weights, such as guns, and a floating dock. Fuel reserves, a pier and base oilers, a reservoir or other means of storing water, a distilling ship, workshops and repair ships, victualling ships, ammunition ships, a torpedo depot, a seaplane station, medical facilities, hospitals, minefields, defences and a garrison would also all be needed.[24]

Operating from an advanced base within 600 miles of Japan would increase lines of communication to some 2,000 miles from Singapore, and 10,000 miles from Britain.[25] The national investment in such a strategy would be wasted if the war plan failed to consider what exactly the fleet was going to do and how the Japanese were going to be defeated. Merely operating, at long range from a base, without any purpose other than the defeat of the enemy, clearly involved unnecessary risks and was not an acceptable strategy. It was essential to have a clear understanding of the strategic aim and to select, at each stage, an objective. Preparation of these plans was impossible without some knowledge of what was to be achieved, the means at one's disposal to achieve this and the conditions under which those means were to be employed.[26]

On 7 March 1921 the first of a series of periodic, biannual meetings of the Flag Officers of the China, East Indies and Australia Stations was held on board HMS *Hawkins* at Penang.[27] Agreement was reached on how each of these stations would function. In peacetime each would remain a separate command, although the Admiralty would remain in overall control, preparing war plans, organizing exercises and manoeuvres, circulating all commanders with the reports of these, authorizing ship visits to foreign ports and intelligence gathering. In time of war, the Commander-in-Chief, East Indies Station, would become Commander-in-Chief, flying his flag ashore, at Singapore.

To a certain extent, the decisions made at the Washington Conference affected British strategic planning regarding the Far East. For a war in the Pacific the Admiralty calculated on having a total naval strength equal to that of Japan, as well as a percentage necessary to compensate for the distance from a major base, and a percentage necessary to keep sufficient forces in home waters – a ratio of three Royal Navy battleships to every two Japanese.[28] The Washington Naval Treaty, by imposing parity with the United States and a 5:3 ratio with the Japanese, actually worked in the Royal Navy's favour. Although the Japanese would have regional superiority this would only be until the Main Fleet arrived at Singapore, a fully functioning naval base. Attacking Singapore would place the Japanese battle fleet at the end of a long line of communications, harassed by cruiser raids from Hong Kong.

Suggested British naval counter-measures therefore centred on the need to show a strong interest in the region through a defended base at Singapore, even though permanently basing a battle fleet in the Far East was out of the question. It was again strongly recommended that the Main Fleet be concentrated in the Mediterranean, ready to move east or west, and that refuelling facilities *en route* to Singapore be improved:

> Adequate fuel reserves in, and on the route to, the Pacific, not only make it possible for our fleet to preserve British interests in the east in the event of war with Japan and to safeguard Hong Kong, Singapore and Australia, but also permit us, in the not unlikely event of hostilities between Japan and America, to retain our fleet in a central position such as the Mediterranean, ready to reinforce either east or west, as strategic necessity may dictate.[29]

With the base at Singapore developed, Britain could delay playing its hand until the last minute. A weaker position, with fewer facilities in place, necessitating the sending of a fleet earlier, might have drawn Britain into the war, as this deployment could have been taken as a provocative action by the Japanese.

As already stated, some thought was also given to the disposition of the battle fleet in the event of war, and the bulk of the Far Eastern fleet was to be drawn from the Mediterranean. The battle cruisers and six cruisers from the Mediterranean Fleet were to steam east to Singapore, to reinforce the China Fleet immediately, followed by a depot ship, flotilla leader, 16 destroyers and six submarines from the Atlantic Fleet, all of which would proceed to reinforce Hong Kong. A further four submarines would proceed to Singapore. All available battleships would be commissioned, along with at least six submarines, three destroyer flotillas, and all aircraft carriers, minelayers and minesweepers in Reserve.

The 1919 plans had proposed that the Mobile Naval Base Defence Organization Ship, the former battleship *Agincourt,* was to be commissioned and proceed to Aden, to prepare anchorages in the Red Sea and Indian Ocean. And although *Agincourt* was scrapped as a part of the Washington settlement, the provision of permanent anchorages would still be needed.

Although still not a strategic plan, this formed the basis of the 1924 memorandum, and preparations and redeployments went ahead. In 1923 a strategic review of the distribution of the fleet took place. Following the conclusion of the Great War, the Admiralty had stationed six battleships in the Mediterranean to support policies there, and retained the rest in the Atlantic Fleet. Stationing more ships there would have cost more:

> Now, however, it has become possible to suggest a gradual redistribution of the battle strength in accordance with the requirements of strategy. As these requirements must mainly depend on the distribution of the naval forces of foreign Powers, these will first be considered.[30]

With the United States Navy's Pacific Fleet unable to develop a naval base in the western Pacific nearer than Pearl Harbor, the Royal Navy, by virtue of its Singapore base, was the only power capable of countering any future aggressive tendencies of Japan, but

> it is neither feasible, economical nor desirable to base our main Naval strength in the Far East today, and mobility must therefore be the keystone of our strategy, by which it is meant that the Fleet must be capable of making as rapid a passage as possible to the threatened area.[31]

The most obvious station was the Mediterranean, and it was proposed to exchange the Atlantic and Mediterranean Fleets in the next few years, stationing the most modern battleships, the five *Queen Elizabeth*-class ships at Malta, along with three light cruiser squadrons, two destroyer flotillas and one aircraft carrier. At Gibraltar there would be a squadron of the Atlantic Fleet, the four *Iron Duke*-class battleships, with one light cruiser squadron, two destroyer flotillas and one aircraft carrier, able to deploy eastwards or westwards, as circumstances demanded.[32] Further re-distributions included stationing the five *R*-class battleships, three battle cruisers, two light cruiser squadrons, four destroyer flotillas, two submarine flotillas and two aircraft carriers as the Atlantic Fleet in home waters, keeping some modern ships to counter any possible European threats. With these distributions, it was believed unlikely that Japan would ever attempt to capture Singapore and unlikely that there would be any war much before 1929. Keeping a battle fleet nearer to the Far East meant that the dry docks in the Mediterranean needed to be enlarged, however, to accommodate a bulged battleship, in addition to developing docks and the necessary oil storage.

Beatty had originally envisaged reviving the pre-war idea of stationing battle cruisers in the Pacific Ocean and wanted a small squadron of the three battlecruisers, an aircraft carrier, six to eight cruisers, 16 destroyers and 21 submarines, the 1923 'Peace Fleet', to be based at Singapore as the nucleus of an Eastern Fleet. This force would have been unable to face a Japanese battle fleet in action but could have formed the nucleus of a larger fleet, covering the oil stocks at Rangoon and Singapore and attacking Japanese trade and convoys. He was, however, advised by the Foreign Office against this as being too provocative, and likely to be seen by Japan as an overtly hostile move by the British. Additionally, the facilities at Singapore were not ready to support any capital ships, even battlecruisers. Instead, the bulk of the forces to be sent eastwards would remain in the Mediterranean, able to reach Singapore within 28 days. From Singapore, the fleet would move to Hong Kong before going to an advanced base, from which to seek out and destroy the Japanese battle fleet.[33]

Tables drawn up for the 1923 Imperial Conference showed the dominating position Singapore was coming to have in naval strategic thinking.[34] Fuel provision could only be solved by long-term planning. But at least a start had been made on finalizing war planning; the disposition and composition of the Main Fleet was settled.

The state of the defences of Hong Kong, and the possible usefulness of the colony were also discussed. Singapore remained the key to any war plan, but Hong Kong could be used as a forward base and point of assembly for operations against Japanese lines of communications or as a base for the battle fleet to operate against the Imperial Japanese Navy, if it could be defended. However, Hong Kong's defences were weak, and Article XIX of the Washington Treaty agreeing to the limitation of fortifications in the western Pacific implied:

Table 2
Oil storage proposals, 1923

	Completed by 31/8/23	In Hand	Authorized, Not Yet Commenced	Total Contemplated
Gibraltar	23,400	23,000	34,200	40,000
Malta	62,700	5,700		102,600
Port Said	16,000			16,000
Port Sudan				24,000
Aden		24,000	24,000	108,000
Colombo		84,000		84,000
Trincomalee		72,000	84,000	156,000
Rangoon		36,000	120,000	420,000
Singapore	228,000	144,000	396,000	1,268,000
Hong Kong	24,000			24,000
Australia	?			420,000
New Zealand	?			27,300

Source: PRO, ADM 116/3438, '1923 Imperial Conference: Admiralty Policy with Regard to Dominion Navies'.

that no new fortifications or naval bases shall be established ... that no measures be taken to increase the existing naval facilities for the repair and maintenance of naval forces, and that no increases shall be made in the coast defences.

This ... does not preclude such repair and replacement of worn out weapons and equipment as is customary in naval and military establishments in time of peace.[35]

In a study of Hong Kong by Lieutenant Commander Stewart, RN, a detailed inventory of the colony's existing defences was made, and judged inadequate: two battalions of infantry, supported by artillerymen and engineers, a local defence force and fixed 9.2-inch and 6-inch coastal batteries, supported field guns, howitzers and mortars.[36] Naval forces included 12 *L*-class submarines in Reserve, the 12 gunboats from the Yangtze and West rivers, and the cruisers of the China Squadron, possibly supplemented, in time of war, by the Australian light cruisers.

Stewart suggested improvements could be made to the defences of Hong Kong, under the guise of renewal, increasing the garrison to between four and 11 brigades with their artillery support, as well as permanently basing fully commissioned submarines and coastal motor boats, armed with torpedoes, at Hong Kong. More importantly and more realistic to achieve, subject to RAF agreement, he suggested two squadrons of reconnaissance aircraft, two squadrons of torpedo bombers, one squadron of seaplanes and one squadron of fighters be based there.[37] This would give Hong Kong the ability to hold out for three to four months, by which time it was expected that the Main Fleet would have arrived.

Stewart concluded that Hong Kong would be lost unless an efficient, fully functioning air base was established, the naval defence flotillas were increased, gun batteries were renewed and re-sited, the military garrison was increased and mines and a minelayer were based at Singapore, ready for immediate transfer to Hong Kong.[38] Improving the defences before Singapore was completed would probably be pushing the spirit of the Washington Treaty too far, however, and would increase the Admiralty's overall costs.

Nevertheless, the First Sea Lord, Admiral Beatty, remained active in keeping up the pressure for the Royal Navy's Far Eastern strategy in the face of government doubts. In his speech at the Lord Mayor's Banquet, on 9 November 1923, he gave a clear assessment of the Admiralty's thinking when he stressed that Singapore was not new, it had always been a naval base and the western Pacific had always been a place for strong British naval forces. The base was being improved and modified to meet the changing world situation.

WAR MEMORANDUM (EASTERN)

Beatty placed responsibility for imperial defence firmly on the government's shoulders, suggesting that if it felt that it could afford to live on the goodwill of Japan, then the Singapore base was a luxury; if the opposite was true, then Singapore was vital. He argued that in a modern war no country was self-supporting, and both Britain and Japan must depend on the outside world for some of the necessities of life. A small disturbance in the flow of these supplies would affect national life and so a navy was necessary to defend trade, by disarming the enemy forces. Risks must be taken in striving for this. War meant fighting, and every opportunity which presented itself had to be taken to test the enemy's defences, find the weak spot and attack with overwhelming force to bring about the destruction or neutralization of the enemy fleet. The offensive powers of a fleet were judged to be great, both economically and strategically, and the presence of fleet in the region, or even the knowledge that such a fleet could deploy to a fully functioning naval base, could have brought pressure to bear on the whole population of Japan.

All of these various strands of planning and thought were drawn together in 1924, in War Memorandum (Eastern). In time of war, the memorandum envisaged being able to move substantial naval forces to a fully functioning naval base, refuelling on the way. These naval forces, the Main Fleet, were to comprise two battleship squadrons, two cruiser squadrons, three destroyer flotillas, two of submarines, two aircraft carriers, three depot ships, one hospital ship and one repair ship from the Mediterranean Fleet, and two battleship squadrons, one battlecruiser squadron, one cruiser squadron, two destroyer flotillas, one submarine flotilla, two aircraft carriers, two minelayers, and one depot ship from the Atlantic Fleet. In addition, a battleship squadron, cruisers as necessary for trade defence, nine destroyer flotillas, three submarine flotillas, one seaplane carrier and six depot ships would be commissioned from the Reserve Fleet.[39] This was still, predominantly, moving warships around. No attempt was made to cover operations to be undertaken after the arrival of the Main Fleet, beyond outlining the three phases of the expected war. As in other Admiralty documents to do with Far Eastern policy, the following point was made:

> No indication has been received from HM Government that diplomatic relations with Japan are anything but satisfactory but it is necessary, as a matter of common precaution, that the administrative and strategical problems of such a war should be carefully investigated with a view to perfecting our arrangements should war with Japan supervene.[40]

Despite the telling phrase 'a matter of common precaution', with Japan being the only potential enemy there had to be careful investigation into the possible causes of war between Great Britain and Japan, of which there seemed many, principally Japan's growing population and need to expand its commercial and political sphere of influence, which it was felt would adversely affect British interests in China.[41]

In other words, war was unlikely, but needed to be planned for, in order to prepare the battle fleet and the strategy to deploy it. The Washington Four Power Treaty and its provision for discussing any controversial matters made any sudden Japanese strikes unlikely, but they could not be ruled out. However, the Admiralty expected to be able to use any period of strained relations to press for the deployment of the battle fleet, although the speed with which the plan could be implemented depended on the government, and on the existence of a battle fleet, fuel supplies and a base.[42]

Singapore was the key to the strategy and cruisers from the China and East Indies squadrons would have to shoulder the burden of defending Singapore until the arrival of the Main Fleet because

> If Singapore were lost the Fleet would be immobilised for want of fuel and would be incapable of relieving the pressure on Hong Kong in time to save it from also falling into the hands of the Japanese.
>
> With Singapore in our possession the situation could be retrieved even if Hong Kong had fallen.
>
> <u>The safety of Singapore must therefore be the keynote of British strategy</u>.[43]

The Royal Navy's numerical superiority over the Imperial Japanese Navy was discounted because of the difficulties posed by operating so far from established bases, which meant that Singapore had to be developed.[44] The memorandum identified three main strands to any future conflict.

Phase I, *The Period before Relief*, would include a Japanese attack on Singapore, which would be held for a period of 42 days, defensive actions being fought by the China, Australian and East Indies squadrons, combined as the Pacific Fleet, with the aim of disrupting Japanese shipping and trade until the arrival of the Main Fleet. Firmly holding Singapore would also reduce the likelihood of attacks on Australia or New Zealand, as there was correspondingly little time for the Japanese to conduct a major combined assault on these Dominions 4,000 miles from their own bases. Phase II, *The Period of Consolidation*, would be the establishment of a trade embargo around the 'Malaya Barrier', the islands stretching from Malaya through the

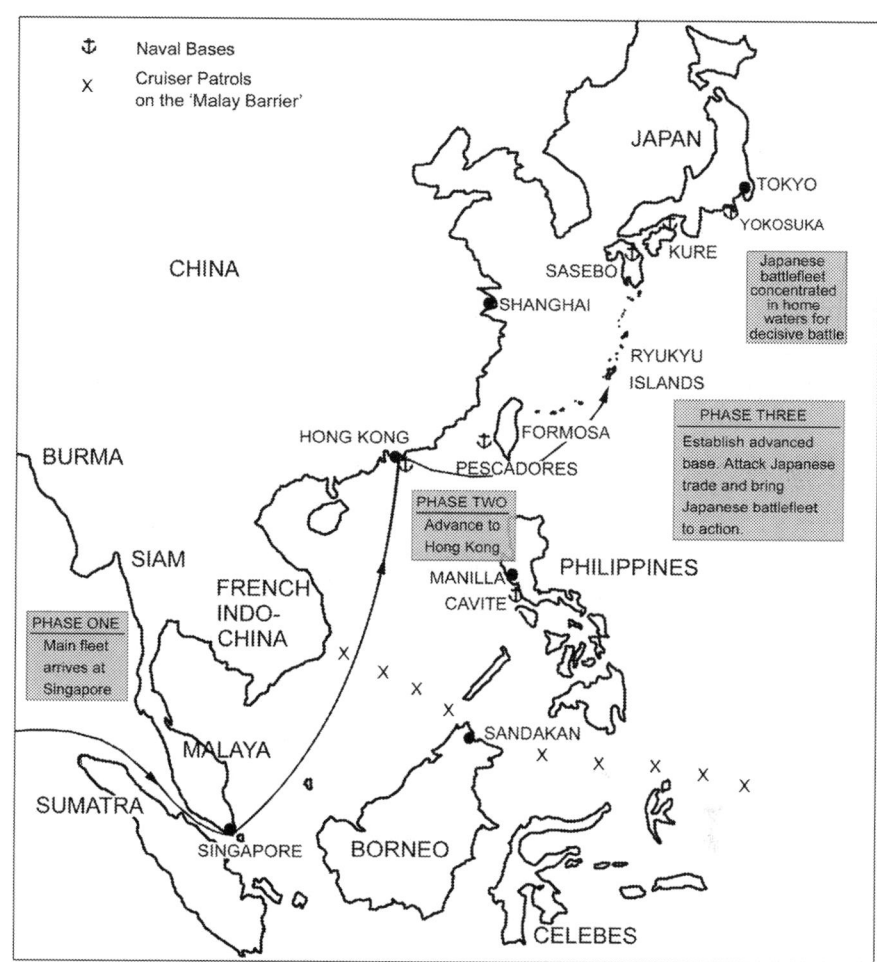

Map 3: War Memorandum (Eastern), 1924–31

Netherlands East Indies to New Guinea and northwards to Japan, with cruisers and armed merchant cruisers operating against Japanese trade, and submarines operating in and around the Sea of Japan. Phase III, *The Period of Advance*, was built around the battle fleet, which would move northwards to Hong Kong, Formosa and the Pescadores Islands, establishing advanced bases for the prolonged actions envisaged in the South China Sea against Japanese trade routes and coastal ports, aimed at forcing the Japanese battle fleet to seek a decisive action.[45]

The nearer the British battle fleet got to Japanese waters, the greater the pressure it was expected to bring to bear against Japanese seaborne trade and economic life, exposing Japan's weakness to economic warfare and forcing the Imperial Japanese Navy to abandon its own strategic aim of enticing the British into Japanese waters for a decisive battle but, instead, forcing it south, to fight on British terms. There were risks attached to this strategy, as it could have placed not only the Japanese, but also the British, at the end of a long line of communication. The advantage of forcing the Japanese to fight near Singapore or an advanced base was that, being further from their dockyard and replenishment facilities, they would inevitably not be able to deploy their entire strength, and would have to husband both fuel and ammunition supplies, as these could be used up quickly, even before the decisive battle, whilst the British, by comparison, could bring the battle fleet, well supplied with fuel and ammunition, to bear, concentrating their available assets and bringing superior force to bear on the Japanese in what was hoped to be a crushing victory.[46]

Echoing Beatty's speeches to the 1921 Imperial Conference and the 1923 Lord Mayor's Banquet, War Memorandum (Eastern) saw the objective of the war as obtaining victory by destroying the Japanese battle fleet, and all planning was directed towards this end. As it was expected that the Japanese might decline battle, the secondary objectives were to secure Singapore and put economic pressure on Japan, making it possible for land forces to capture Japanese possessions and force the Imperial Japanese Navy to take risks with its battle fleet.

Previous studies had already indicated the problems involved in moving a large fleet eastwards, and although there may have been sufficient oil to deploy the fleet to Singapore, reserves needed to be built up to make the base self-sufficient, and not dependent on vulnerable re-supply convoys. Secret bases, Bases *M* (Kamaran Bay, in the Red Sea), *T* (Addu Atoll) and *N* (Nancowry Harbour), were selected, never to be referred to by name.[47] Their existence and safe preservation was vital for the passage of the Main Fleet. It was also anticipated that by 1931 the full authorized reserve of 1,250,000 tons of fuel oil would be stored at Singapore, which would allow a battle fleet to conduct operations against Japan.[48]

Until Singapore could be fully developed as a naval repair port, any fleet in the East would be dependent on the commercial facilities at Singapore itself, Colombo and Calcutta, whilst any damaged battleships would have to return to Durban or to Malta for repair, as these were the only places with docks able to take bulged battleships. As an interim, it was intended to position a large floating dock and a modern fleet repair ship, *Assistance*, in the Old Strait at Singapore, although the battle fleet was to be as self-dependent as possible.[49]

It was considered unlikely that the Japanese would use submarines against British merchant shipping and attacks by surface raiders were judged to be much more likely. This meant emphasizing the surface defence of convoys and shipping, not anti-submarine warfare.[50] If Singapore was secure, and could be held, and local sea superiority secured, the Japanese task would be made much more difficult. Japanese success in denying Singapore to the British would be dependent on the level of force it was prepared to devote to such an operation. Japan's ultimate success would have been to delay the arrival of the Main Fleet, and to draw the British northwards, using submarines and light craft to wear down the fleet before a battle in Japanese waters.

Following on from the 1921 meeting, a further Commanders-in-Chief meeting at Penang had been planned for 1923, and in a draft paper an attempt was made to define responsibilities for the defence of the region more clearly. This was seen to be especially important for the Admiralty if they were to maintain overall, strategic control, although it was presented to the Dominions as being necessary to reduce Britain's burden:

> In view of recent reductions of the British Fleet and the possible further expansion of Dominion Fleets it is considered that they [the requirements for co-operation between British and Dominion Navies] now require some elaboration and amendment... The British Empire having agreed to a One Power Standard is thus faced with the problem of protecting its vast maritime interests with a navy no greater than that of any other naval Power. Furthermore the British Empire differs from other Powers in that its naval quota is divisible between Great Britain and the Dominions.[51]

At the time of the writing of the draft paper, Britain was responsible for the maintenance of the major portion of imperial defence, but it was reasonable to assume some changes would be made, possibly including a change to the funding of the Empire's One Power Standard, which might be made up with Britain providing 50 per cent of the total and the Dominions the other 50 per cent. However it was to be arranged, there was a grave danger that the collective efficiency might fall short of what was required to safeguard the defence of the Empire unless the navies were trained to act as one in time of war and unless a clear war plan, with precise strategic objectives, existed.

The meeting at Penang was postponed until after the Committee of Imperial Defence's memorandum, 'The Washington Conference and its Effects upon Empire Naval Policy and Co-operation' was digested, and the Plans Division presumed that it would therefore not take place until after the forthcoming Imperial Conference. Captain Pound, DCNS, suggested 1924 would be a good time to meet, as by then the new Flag Officer, China, would have had time to familiarize himself with the local situation, whilst Richmond, Commander-in-Chief, East Indies, who was being relieved in 1925, had made a special study of the problems involved in an Eastern war, and the conference could find his opinions and knowledge extremely useful. Australia was again rejected as the venue for the conference as it was felt that it would take the various flag officers too far away from their respective stations and Admiral Everett, the Flag Officer, Australia and New Zealand, believed that the recent visit of the Special Service Squadron (the battle cruisers *Hood* and *Repulse*, and the 1st Light Cruiser Squadron, *Delhi*, *Danae*, *Dauntless* and *Dragon*) had already done a great deal towards increasing awareness of the navy in the Dominions. As the development of the Singapore base had just been postponed, it was felt further inadvisable to hold the conference in Sydney as this might make the Australians feel that a base in Australia was to be developed as an alternative.[52]

When War Memorandum (Eastern) was eventually discussed at the 1925 Flag Officers' Conference at Penang, it was agreed that, in the event of war,

> it is proposed that 2 Royal Australian Navy cruisers, 5 Royal Australian Navy destroyers and the 2 New Zealand cruisers should first assemble at Port Darwin ready to act in accordance with the orders of the Commander in Chief, Eastern Forces, to assist in the attainment of the 'Object', *viz.*, the security of Singapore and the passage of the Main Fleet.
>
> This will leave the New Zealand and Australian trade routes temporarily rather weakly defended but it is undoubtedly correct strategy to concentrate on the Main Object and if the Governments of Australia and New Zealand are prepared to acquiesce in this plan, there is every reason for the Admiralty to support it.[53]

Imperial naval forces were thus to concentrate on the Straits of Malacca to ensure the security of Singapore before the arrival of the Main Fleet and it was originally intended to have the cruisers from the battle fleet protecting the lines of communication between Singapore, Hong Kong and, once established, the advanced base. The Conference now considered this impossible and, instead, it recommended a force of four cruisers based on North Borneo to do this, with cruiser patrols between North Borneo and the Philippines, and between Java and Port Darwin. To do this, more modern ships would have to be withdrawn from other stations, being replaced by vessels from Reserve, or by armed merchant cruisers. Three convoys of troop reinforcements, from Madras, Calcutta and Rangoon, would reinforce the garrison protected by concentrations of ships in the Straits of Malacca, and patrols would be in the Sunda Strait, so that

> any attempt on the part of Japanese Battle Cruisers to penetrate into the Bay of Bengal via the Malacca Straits will become known in time to divert or to recall the transports.
>
> A possible, but very real danger is that the Japanese Battlecruisers will utilize the Straits of Sunda and it is strongly recommended therefore, that a patrol should be maintained in these waters from the receipt of the warning telegram.[54]

The proposed force for the Malacca Straits was to be the single battlecruiser squadron, one cruiser squadron and a destroyer flotilla, supporting a cruiser squadron, an aircraft carrier and two destroyer flotillas, essentially the fleet Beatty had wanted to base in the region in 1923, but which would now be despatched from the Mediterranean Sea and supported by aircraft from Singapore. Later, in another confirmation of the Royal Navy's commitment to air power, the aircraft carrier force would be sent on ahead, to reach the Straits 48 hours ahead of the Main Fleet and provide air cover for their passage into Singapore.

It was also agreed that submarines were needed to patrol the Sunda Strait on receipt of the war-warning telegram and that it was also vital to control the navigational lights in the Malacca and Singapore Straits. These could easily be extinguished in wartime, despite the isolated nature of some of them and the difficulties of communication with them. It would not be possible to prevent entirely navigation of these straits, as the lights on the Dutch side of the Singapore Straits were sufficient for navigation in clear weather.

In the Straits of Malacca, southward of Aroa Island, eastward navigation would be difficult without the British lights, but it would not be absolutely impossible; it would be necessary to proceed slowly, making use of soundings, so it would not be possible to complete the passage from Aroa Island to Singapore (220 miles) in one night; but anchorage could be found on the Dutch coast out of visual range of the regular steamer track.[55]

As cruiser patrols were vital, it was proposed to develop a cruiser base, with a 5,000-ton fuel storage at Port Darwin.[56] Other recommendations were to despatch one of the 15-inch gunned monitors, either *Terror* or *Erebus*, to bolster the defences of Singapore and Hong Kong, rather than be employed in operations such as capture of the Pescadores Islands in the event of war.

The question of whether Hong Kong had a role, and whether it could be defended, arose again. The Conference members noted that whilst the loss of Singapore would be a serious blow to British strategy, so too would that of Hong Kong, and insufficient weight had been given to its defence.[57] The General Officer Commanding, Hong Kong, believed that if the naval defence provision was increased, Hong Kong could actually hold out with a small garrison, a much more optimistic assessment than the War Office's. The Conference therefore recommended:

> As a preliminary step it is proposed to hold a Naval Staff conference to examine all possibility, within the limits of the Washington Treaty, for strengthening the resistance to attack which could be made by Hong Kong against combined Naval and Military forces.[58]

It was proposed to have a total of seven submarines, four sloops, ten gunboats and one submarine depot ship at Hong Kong, eight cruisers, four sloops, 12 minesweepers, five submarines and one depot ship at Singapore, four cruisers and eight destroyers at Port Darwin, Australia, with a further two cruisers and eight destroyers further south, with sloops at Nancowry and Basra. An extra destroyer flotilla was recommended for anti-submarine work, although it was not considered possible to reinforce the China Station with more destroyers before 1929, partly because of adverse political opinion.[59]

Further suggestions were also made by the flag officers at Penang, such as denying Japan a supply of oil by purchasing the entire Netherlands East Indies oil output. Not only would this release cruisers from hunting down Japanese tankers trading with the Dutch East Indies, allowing them to concentrate on the more delicate problem of stopping the flow of oil from the Americas to Japan, but further reserves of oil, refined at Rangoon or the Dutch refineries around Palembang, in Sumatra, would be available to the Royal Navy.

WAR MEMORANDUM (EASTERN)

Richmond, Commander-in-Chief, East Indies Station, identified what he saw as the flaws in the plan, the principal one being that it was not a war plan, but still a plan of naval movements. Early in 1924 he had complained to Lord Haldane, Chairman of the Committee of Imperial Defence, that the Memorandum was failing to address the question as to what the battle fleet was going to do once it arrived at Singapore, and that planning needed to concentrate on how Britain was going to use its military and economic strength to defeat Japan. The following year, in a critique to the Plans Division, he reiterated that War Memorandum (Eastern) was a plan of naval movements only, presuming a decisive battle and seeing a war with Japan purely as a struggle at sea.[60]

NOTES

1. Rear Admiral Bradley A. Fiske, USN, *The Navy as a Fighting Machine*, ed., W. Hughes (Naval Institute Press, Annapolis, MD, 1988), pp. 141–2.
2. PRO, ADM 116/3104, 'Imperial Naval Defence'; Tracy, *The Collective Naval Defence*, p. 252.
3. PRO, ADM 1/8/8570/287, 'British Imperial Naval Bases in the Pacific, 28 April 1919', p. 3.
4. PRO, CAB 16/37/1, 'CID Sub Committee on the Question of the Capital Ship in the Navy, Vol. I, 11 January 1921', p. 601.
5. Ibid., p. 596.
6. Ibid., p. 116.
7. Ong Chit Chung, *Operation Matador: Britain's War Plans against the Japanese 1918–1941* (Times Academic Press, Singapore, 1997), p. 29.
8. Ibid., p. 31.
9. I. Cowman, *Dominion or Decline: Anglo-American Naval Relations in the Pacific, 1937–1941* (Berg, Oxford, 1996), p. 16.
10. ADM 116/3102, 'Oil Fuel Reserves; Recommendation by VA Sir Osmond de B Brock, DCNS, 24 May 1921'; Tracy, *Collective Naval Defence*, p. 283.
11. Ibid., p. 283.
12. Ibid., p. 284.
13. Ibid., p. 285.
14. *Naval Historical Branch, London*, W.J.R. Gardner, to author, 22 August 1997.
15. CID 145-C, CAB 5/4, 'Reserves of Oil Fuel – Memorandum By Lord Lee of Fareham, First Lord of the Admiralty, 21 June 1921'; Tracy, *Collective Naval Defence*, p. 293.
16. CID 143-C, CAB 5/4, 'Singapore, Development of as a Naval Base. Memorandum by the Overseas Sub Committee of the Committee of Imperial Defence, 7 June 1921'; Tracy, *Collective Naval Defence*, p. 287.
17. Ibid., p. 288.
18. CAB 32/2, 'Beatty's Address to the Imperial Conference', 4 July 1921; B. Ranft (ed.), *The Beatty Papers, Vol. II, 1916–1927* (Navy Records Society, London, 1993), p. 177.
19. Ibid.
20. P. Haggie, *Britannia at Bay* (Oxford University Press, Oxford, 1981), p. 10.

21. Ranft, *The Beatty Papers*, p. 177.
22. Ibid., p. 140.
23. NMM, Richmond Papers, RIC 10/5b, 'Naval Lines of Communication', Captain G. Hamilton, RN, 1 April 1921.
24. NMM, Richmond Papers, RIC 10/5b, 'Strategy 2, Autumn Session 1921'.
25. NMM, Richmond Papers, RIC 11/1, 'Supply of Fleets and Bases', Captain N.F. Lawrence, RN, 22 March 1922.
26. Fiske, *The Navy as a Fighting Machine*, p. 152.
27. PRO, ADM 116/3167, 'Pacific War Plans and Naval Problems 1921–1925'. Significantly, it was felt undesirable to agree to an Australian request to hold it at Sydney as it was felt that not only would this give colour to the deeply rooted idea that the Australasian ports would become the centre of a Pacific strategy, but also that the Commanders-in-Chief might have found themselves drawn into discussing this, a political topic.
28. PRO, ADM 116/3445, 'Washington Conference on Limitation of Armaments and Far Eastern Policy 1921–22, Vol. 1, 5 October 1921'.
29. PRO, ADM 1/8948, 'The Naval Situation of the British Empire in the Event of War between Japan and the US of America, 1919', p. 8.
30. PRO, ADM 116/3195, 'Redistribution of the Fleet to Meet Changes in the World Political Situation and a Possible War Threat in the Middle and Far East, 1923', p. 12.
31. Ibid., p. 14.
32. Ibid., 'Approximate Cost of Upkeep of Battleship of Queen Elizabeth Class in 1922–23 on Home and Mediterranean Station', p. 18. It would be more expensive keeping capital ships on foreign stations – £410,200 to maintain the modern ships in home waters, and £421,700 in the Mediterranean, an increase of £11,000, but it would still be cheaper than permanent deployment to Singapore.
33. Ranft, *The Beatty Papers*, pp. 139–40.
34. PRO, ADM 116/3438, '1923 Imperial Conference: Admiralty Policy with Regard to Dominion Navies'.
35. PRO, ADM 203/47, 'The Possibility of Losing Hong Kong in a War with Japan in 1925', Lieutenant Commander R.R. Stewart, RN.
36. Ibid.
37. Ibid.
38. Ibid.
39. PRO, ADM 116/3125, 'War Memorandum (Eastern), August 1924'.
40. Ibid., 'Guidance for War against Japan. Covering Letter to War Memorandum (Eastern), August 1924', p. 3.
41. Ibid., p. 4.
42. Ibid., p. 16.
43. Ibid., p. 9.
44. Ibid., p. 5.
45. Ibid., pp. 10–11.
46. E. Grove, *The Future of Sea Power* (Routledge, London, 1990), p. 18. The reverse was also true. The British were also operating at the end of a long line of communication.
47. PRO, ADM 116/3125, 'War Memorandum (Eastern)', p. 5.
48. Ibid., p. 17.
49. Ibid., p. 18.
50. Ibid., pp. 18–19.

WAR MEMORANDUM (EASTERN)

51. PRO, ADM 116/2274, 'Empire Naval Cooperation. Draft of Proposed Paper to be Laid before Imperial Conference, 1923, as a memorandum'.
52. Ibid.
53. PRO, ADM 116/3121, 'Flag Officers at Singapore Conference', 1925, p. 12.
54. Ibid., pp. 13–14.
55. Ibid., p. 66.
56. Ibid., p. 43.
57. Ibid., p. 18.
58. Ibid., p. 27.
59. Ibid., p. 23.
60. C. Bell, 'How Are We Going to Make War? Admiral Sir Herbert Richmond and British Far Eastern War Plans', *Journal of Strategic Studies*, Vol. 2, No. 3, 1997, p. 128.

3

Admiral Richmond and War Memorandum (Eastern)

Richmond had already devoted a lot of time and mental energy to the ideas associated with fighting Japan and was very sceptical about the plans drawn up by the Admiralty, and discussed at Penang in 1925. Between 1921 and 1923 he had been the Commandant of the Royal Navy's Staff College at Greenwich, where the strategic problems of a Far Eastern war had been discussed, and in one of his Commandant's Lectures, Captain Richmond had presented an alternative scenario to the Admiralty's hypothetical attack on Singapore as a way of showing that the movement of battle fleets and a decisive battle was not, in itself, a strategy.[1]

Instead, he suggested that a war could be started following a hypothetical Japanese invasion of Fukien province; in China, Britain had gone to war over the way this invasion threatened British interests, and Japan, in turn, had counter-attacked British trade with cruisers and submarines. The British fleet had, as expected, arrived in the East and fought a fleet action, which the Japanese had lost, exactly as the Admiralty expected. But the British victory had not severed the lines of supply to Fukien, nor stopped Japanese attacks on British trade, nor even begun to put any economic or military pressure on Japan.

What Richmond believed was lacking in War Memorandum (Eastern), and what the navy actually needed, was a clear set of strategic guidelines from the government, at the very start of a campaign, as to the proper purpose of Britain's war plans and strategy. In his hypothetical case, Richmond showed that whilst the aim may have been to bring the enemy to terms after a decisive battle, the objective should have been to exert economic pressure on Japan's national life, through economic warfare. The aim of the war planner and strategist was to outline the ways and methods by which this would be achieved and, as War Memorandum (Eastern) was refined, more thought was put into this aspect.

Richmond identified what he felt were the two key points to consider in any set of war aims. First, Japan was, in fact, difficult to blockade, producing 95 per cent of its own food. With a long Pacific coastline, trade with Europe could only be stopped by cruisers operating from Singapore or on the Panama Canal trade routes to Europe. However, Japan's trade with the United States would be difficult to prevent because of diplomatic complications, hence the need for a friendly United States.[2] Richmond believed that Singapore would be relatively safe from attack and the defence of India was the main consideration for any Far Eastern strategy, not the relief of Singapore and an advance up the Chinese coast to Japanese waters, in search of a decisive surface engagement. A naval victory would not necessarily end the war and Singapore and Hong Kong remained important to this strategy only in so far as they could act as bases from which warships could sail to defend trade in the Indian Ocean. He particularly warned his students against a strategy of generalizations, which had distinct limitations, referring back to the Great War to illustrate his point:

> We did indeed, before the late war, talk of 'the offensive', of 'seeking out and destroying the enemy', and of 'fighting decisive actions' in which the only aim was to be the annihilation of the enemy. But in practice we did not do these things. As generalizations they were admirable; but if you examine – as we shall – our plans for war, and our strategy and tactics you will find that they are not universally informed with the particular spirit of the offensive, that we did not make any attempt to 'seek out' the enemy.[3]

As close blockade of enemy harbours was impossible because of the threat of attack by air or submarines at any time, or by destroyers at night, the Royal Navy would have to try to draw the Japanese battle fleet away from Japanese home waters by a campaign against Japanese trade with China and the Netherlands East Indies, carried out by cruisers and supported by battleships. Such actions would inevitably escalate as each side committed heavier forces to screen their own light forces, making battle fleet action more probable.[4] But decisive battle was not going to be enough to defeat the Japanese, and a long-drawn-out campaign against Japanese trade, with all of the difficulties that would entail, would be necessary. Some way had to be found of isolating the Japanese and breaking their will to fight, and in such a war it was not going to be just Japan that would find itself placed under an economic strain; Britain would suffer almost as much, and at the College there seemed to be some clear understanding of the problems associated with a Far Eastern war. Naval strategy was more involved than merely concentrating ships in a given area, locating and giving battle to an enemy.[5] Other lectures considered the logistical side of strategy, an important consideration in the movement of any large group of ships from British waters:

> Doubtless, many of us, during the strategy lectures, have in our minds transferred the British fleet to Singapore. It may be of interest to know that under existing circumstances a period of about three months would elapse from the time the Fleet left for the Far East before it found itself in a refuelled condition at Singapore. It could then only hope for irregular supplies and could certainly not operate continuously for a month. Further, to get the fleet to the East would mean the taking up of practically the whole of the tanker tonnage under British control, which would, of course, completely disorganize the supply of oil for commercial purposes throughout the Empire.[6]

These were serious considerations for Britain if it was going to go to war with Japan as, obviously, the longer it was at war, the greater would be the effect on industry and domestic life, with either fuel rationing or increased taxation to pay for tanker charters.

Richmond even challenged the assumption that the best means of achieving Britain's objectives was through an economic blockade, citing the experiences of Britain and Germany in the Great War, which illustrated how long countries could keep fighting even with severe disruption to their trade:

> Our power of exercising economic pressure and the resulting effects we are preparing to make are greatly overestimated in the Memorandum. A war of economic exhaustion is proverbially long, and Japan is peculiarly well situated to hold out against it. Allied to military action, as our economic efforts in past wars have nearly always been, it is an instrument of far reaching effects, but by itself, economic pressure is most unlikely, *pace* experience, to produce surrender.[7]

Richmond did not believe that the Japanese would be any easier than the Germans to break economically, and thought that a purely naval strategic plan, which ignored the contributions of both the army and RAF, was just not sufficient. The objectives of the war would have to be considered carefully, as these would shape how the war would be fought. The success of Japanese policy might not rest upon naval operations at all, but upon the operations of the army in China. From

Richmond's analysis, the shortcomings of a purely naval war plan, with the army and RAF in subsidiary roles, were all too obvious:

> It appears to me that this arises from the manner in which the problem is approached. The cause of the war is not related to the object, nor the object to the operations ... Attempts to reduce Japan by economic pressure, in a war arising out of the cause indicated, would be misdirected. Instead of dissipating our forces all over the world to intercept Japanese trade, we should concentrate our naval and military forces with the object of dealing with the situation in China...
>
> ... We should have the choice of two causes [sic] of action – defeating the enemy's army in the territory under dispute, and defeat by cutting its communications – Assault or Investment.[8]

The memorandum had as the immediate objective the destruction or disablement of the enemy fleet; this was traditionally what battle fleets existed to do. However, Richmond did not believe that the appearance of the Main Fleet at Singapore was going to lead automatically to a fleet action, and recent war experiences showed just how difficult it was to bring an enemy to battle. Richmond wrote:

> I know of no instances in the history of war at sea in which it has proved possible to force an unwilling enemy to sea with the prospect of an action ... Nor do I recollect an instance in which the main strategy has been specifically directed towards that aim ... No one has yet discovered a means of forcing an inferior enemy to sea except by military attack on his base.[9]

In this case, the British battle fleet would be needed to screen the cruisers, and some of the cruisers engaged in attacking trade would be needed for fleet work. The battle fleet, or squadrons of it, would have to be used to attack lines of communication, to force the Japanese fleet to accept action, which placed the British battle fleet at greater risk by dispersing it. Richmond pointed out that if the presence of a fleet which declined to be drawn into action, except on terms favourable to itself, prevented successful attacks on maritime communications, it was by no means clear how the attack upon them was to be achieved, and that this was a weak point of the memorandum; to him,

> The plan depends upon the reasoning which runs in a vicious circle.
>
> (a) We are going to force Japan to surrender by cutting off her essential supplies.

(b) We cannot cut off her essential supplies until we defeat her fleet.
(c) We cannot defeat her fleet if it will not come out to fight.
(d) We shall force it to come out and fight by cutting off its essential supplies.

which brings us back again to (b) and to (c) and (d) in succession.[10]

In such an instance it seemed obvious that the memorandum needed a lot of work on it before it became the serious, strategical document Richmond felt was vital.

There were many other flaws in War Memorandum (Eastern), that Richmond identified. It was, for example, stated that the success of the Japanese plan depended on immunity from the British fleet, and that Japan's primary object would be to gain time by delaying the passage of the British fleet. But Richmond also pointed out that

> delaying the fleet is not the only way of attaining immunity ... there is another – namely, rendering it impossible for the fleet to act effectively when it does arrive: or, to put it more shortly, 'disabling' the fleet. By 'effectively', I mean producing the effect of making Britain's efforts fruitless ... Delaying the fleet would delay the date of interference, but it would not prevent eventual interference. The Japanese plan of campaign will be spoilt if the British fleet arrives and is in a condition to act ... I do not attempt to forecast what the result of an examination of this problem would be: but it appears to me that her primary naval and military object will more accurately lie in one of these operations than in delaying the passage of the fleet to the East...[11]

The British Main Fleet could not arrive for about four or five weeks and, during this time, Japan could probably try for some sort of decisive blow, again probably through attacking Hong Kong. It was believed that a decisive blow against Singapore was clearly beyond Japan's capability in the time judged to be available to it and that it would therefore aim to make itself secure against any interference by the battle fleet. Richmond had, in fact, organized an exercise which seemed to illustrate the difficulties the Japanese would face in invading Singapore.

'The Invasion of Salsette Island', a combined naval and military exercise carried out on Salsette Island, Bombay, organized in 1924 by Vice Admiral Richmond and the Commandant of the Staff College, Quetta, Major-General Boyd, seemed to add validity to the belief that Singapore was a difficult objective to capture; this argument added weight to both the Admiralty's claims that the defences should be as strong as possible, as well as to Richmond's, that Singapore would not be a Japanese objective.[12]

The exercise took place between 30 November and 9 December 1924, simulating an amphibious landing on Singapore Island by Japanese forces. Salsette Island, off Bombay, was substituted for Singapore, with a garrison composed of four infantry battalions, plus gunners and auxiliary troops. The island had to be taken in 35 days, the calculated time it would take the British Main Fleet to arrive, and the results of the exercise are worth considering as it was influential in determining much of Britain's subsequent strategy in the region.

A Japanese naval force of three battleships, two battlecruisers, one aircraft carrier, 12 cruisers, 15 destroyers, one oiler and six fast transports was judged to have sailed from the Pescadores Islands to capture Port Swettenham, on the west coast of the Malayan Peninsula, and from there to act to stop troop reinforcements arriving from India. Significantly for the exercise, and possibly for all subsequent considerations of the defence of Singapore,

> Owing to the threat of aircraft the Plan included the preparation of an aerodrome in the Mersing area. The capture of the site taking place 12 days before the arrival of the main Convoy at Salsette. This was considered necessary for the operations of the aircraft, the 4 aircraft carriers not being able to supply sufficient force. The air officers assumed that sufficient air supremacy would be obtained in two days, operating from the aerodrome at Mersing.[13]

This was an extremely important point, revealing opinions on the usefulness of naval air power in support of land operations. Japanese aircraft carriers (in 1924 the small *Hosho* and a seaplane carrier), did not provide sufficient aircraft to support the landings, or to obtain and keep air superiority over the island. Even projecting ahead, to the end of the decade when Singapore's base facilities and defences would have been completed, as the exercise appeared to do, it would have to be assumed that the Japanese naval commander would not want to risk his aircraft carriers in the confined waters of the Malacca Straits, with his fleet at the end of a long line of communications and with other units, probably the battlecruisers, operating in the Indian Ocean against convoys of troop reinforcements from India. Singapore was, after all, nearly 3,000 miles away from Japan and nearly 2,000 miles from the Pescadores Islands, Japan's nearest naval base. This was a very influential point for the Admiralty, and was one of the reasons why they pushed for a completed base and defences; completing the defences would make Singapore too dangerous to attack in a single operation. A staged attack on Singapore was judged to be more likely, building up bases and probably attacking Hong Kong as a part of the build-up. It was for this very reason that Richmond suggested ending Phase 1 at Hong Kong, and not Singapore. A strong Hong Kong, on the flanks of the Japanese lines of communication, would act as a further defence for Singapore, and thus India, as the time that the Japanese would spend in taking Hong Kong would allow the naval and military reinforcements to reach Singapore.[14] Nevertheless, for the purposes of the exercise, Port Swettenham was judged to be captured five days after war had been declared, and Japanese forces were thus in a position to threaten troop convoys from India. The rest of the invasion plan would then be implemented.

The force detailed for the capture of the airfield site at Mersing comprised two battle cruisers, three aircraft carriers, six cruisers, 30 destroyers, ten submarines and 23 transports, with the landing to be effected ten days after the declaration of war. A further 12 days after this, the main attack on Singapore (Salsette) Island would be expected by military forces carried in 91 transports, escorted by three battleships, seven cruisers, 36 destroyers and ten submarines. Virtually the entire Imperial Japanese Navy and a substantial portion of its merchant marine would be expected to be used in this massive undertaking, and failure would have serious maritime consequences:

> .The main force for the capture of Salsette and Bombay consisted of one division and Auxiliary troops ... As the country was very low lying direct fire from ships was possible on the flanks of the enemy's positions. Nine days were supposed to be occupied in landing stores and laying out the base before the attack on the outer defences could take place.
>
> The capture of the outer defences was calculated for the 12th day after the landing, or two days before the British Main Fleet was expected to arrive in the vicinity of the Singapore Strait.[15]

Everything had to be carried out to a rigorous timetable, and unsurprisingly, the Japanese were judged to have failed in their objective. Instead of a speedy, night landing, quickly establishing a perimeter to secure the beachhead, the assault force ran into difficulties. Among the problems that they encountered were the lack of suitable reconnaissance of landing sites, the breakdown of towing boats, and, a lack of specialized landing craft, which meant the assault force had to row towards the beach against an offshore wind and had to face the difficulties in landing in a strong surf, which pushed the landing boats broadside on to the beach. Other difficulties included the lack of amphibious tanks, as armour support for the troops on the ground as soon as possible was seen to be vital. The amount of motor transport generally – some 1,600 vehicles for this exercise – suggested that a means of landing them had to be devised, either a specialist ship, served by ferries and lighters to the beach, or a sectional jetty, which shallow-draft ships could then moor against.[16]

A means of controlling naval gunfire support was also needed, as was a strong anti-aircraft battery, to protect the beachhead from the defenders' air attacks. In many ways it was the air aspect that had the greatest influence on the failure to capture the island, and possibly therefore on subsequent views regarding Singapore, and it led Richmond to make the following comment:

> The fear shown by the Military authorities of the Air menace was most marked, and was, in my opinion, the cause of the 'failure' of the expedition. Although they fully realised that 'time' was the most important factor, yet 12 days were spent in erecting an aerodrome so that 'air supremacy' should be obtained before the main landing took place. This straining after air supremacy robbed the expedition of any chance of completing its operation within the limit of time available. I consider that the risk of landing without obtaining that supremacy would have been justified, and that the delay caused by waiting for it was not justified.[17]

Richmond was acutely aware of the naval side of operations, but felt that the risks incurred in landing without complete air supremacy were justified: failing to take these risks had left only a fortnight to completely subjugate the objective, Singapore.

He may or may not have not been correct in his conclusions, but only war could offer definitive proof. However, with circumstances as they were, a bombardment and raid on a fully defended Singapore was unlikely to cause much damage, whilst great risks would be taken, exposing the Japanese battleships and aircraft carriers to land-based air attacks and their warships to shore batteries.

A battle fleet was dependent on supplies of fuel oil, food and ammunition and, with the problems inherent in supporting an invasion fleet 2,000 miles away from a main base, a Japanese invasion of a fully developed and fortified Singapore could be expected to fail, with the British fleet arriving before the island was secured and forcing a naval battle on a Japanese fleet. Unlike the Japanese, the British would have replenished at Nancowry Harbour before the final approach, although some ships, especially the destroyers, would be starting to become low on fuel. So, although a raid was a possibility, distance and the risks involved all suggested a full-scale invasion was likely to fail. This seemed to satisfy the

Admiralty; an adequately defended Singapore was virtually immune to attack, and the Main Fleet would be able to operate from there with ease.

Staff College Lectures in 1924, concentrating on the mobility of fleets, re-emphasized the Japanese difficulties by pointing out that a modern battle fleet was a weapon of comparatively short range, and the larger the fleet the greater the difficulty of operating at increasing ranges, whilst demanding constant resupply of ammunition, fuel and food. But this was also true for the British battle fleet operating in the Far East:

> Without careful preparation, good staff work and an exact appreciation of supply and maintenance, such operations will be doomed to failure ... it is not until you remove the Fleet 5,000 miles or so from its Home Country that you really see the problem in its right perspective.
>
> In other words, long lines of communications and a lack of resources – both in stores, fuel, docking and repair facilities and personnel – in the theatre of war may so weaken a numerically superior force that when the tactical phase is reached it may actually prove to be inferior in numbers and fighting capacity to a fleet operation near its own coast.[18]

Clearly, this was another argument in favour of developing Singapore as a first-class naval base, a bastion on which to draw the Japanese in fruitless attacks.

Leaving some of the fleet in European waters – something the writers of War Memorandum (Eastern) saw as a vital part of overall naval strategy if security and influence was to be maintained – was something of which Richmond was critical, as this meant that the margin of superiority over the Japanese would be small compared with the magnitude of the issue. If all available battleships were sent east, then in strictly numerical terms the British fleet would outgun the Japanese, if the circumstances were right.

> Further, if war should come suddenly, it is not improbable that one, or two, of the British ships will be absent for docking or repairs: and the *Tiger*, even if she accompanied the Fleet, would be newly commissioned, as she was in 1915. If that should be, the resulting British figures would be either 100 (one battleship absent) or 92 (two absent): the value of the 8 guns of a newly commissioned *Tiger* must be a matter of opinion. This margin is a very narrow one with which to begin a war with Japan with the certainty that we cannot at once send all the force necessary to carry it to a successful conclusion, and

the distinct possibility, as indicated by the words 'at a later date' that we may have to keep a force at home in case difficulties should arise.[19]

Richmond believed that as the main object of the British fleet was to render the opposition of the Japanese fleet ineffective as quickly as possible, either by destroying or disabling it, then not only every available battleship, but also all available cruisers and submarines, along with quantities of mines, should be despatched eastwards, forcing the Imperial Japanese Navy on to the defensive. Tactical doctrine would also have to be developed to ensure a quick, decisive victory. The success of British strategic plans would be dependent on the success of its battle fleet tactics.

Richmond made some lengthy criticisms of the Admiralty plan. As Commander-in-Chief, East Indies, Richmond believed that the naval forces at his disposal would be better used before the arrival of the Main Fleet in escorting reinforcements to Singapore from India, rather than on attacks on the Japanese invasion force from Hong Kong, masked as they would be by the Japanese battle fleet.

It was recognized from the start that there were problems with Phase III, and Richmond's critique was valuable. Admiral Edmund Slade, a former Director of Plans and now involved with the writing of War Memorandum (Eastern), believed that the building of a naval base at Singapore was directed at stopping Japanese expansion towards India, although he did not think that they would be able to launch an attack against it because of the distances and logistical difficulties involved.

A modern battle fleet was dependent on supplies (which could be calculated for), ammunition (which could be expended in a few hours and was therefore difficult to calculate with accuracy) and communications (which could be disrupted by enemy cruiser raids) and the Japanese would be presenting themselves with a major undertaking in attacking Singapore. The distance between Singapore and the nearest Japanese naval bases in Japan was 2,700 to 2,900 miles by the most direct route, an additional 4,000 by the route east of the Philippines and Borneo. By comparison, Singapore to Trincomalee was 1,500 miles, to Aden 3,600 miles and to the Mediterranean, 5,000 miles. So the Japanese would only have a start of about one week over vessels leaving the Mediterranean Sea. This meant that

> So long as we have at call any vessels capable of attacking their ships within a short time after they become committed to the operation they are not likely to risk finding their vessels depleted of ammunition at the critical moment for the sake of carrying out a raid, that without troops cannot be decisive in any way. A British squadron within

striking distance could be quite sufficient to prevent any action of this sort.[20]

Richmond, however, pointed out to Slade that the war planners had neglected the fact that the Japanese could set out from the Pescadores, only 1,800 miles away.

Attacks on Singapore by ships alone could largely be ruled out, unless the object was to raid the base, which would not stop a British fleet from operating from it. Even destroyers, submarines and torpedo bombers could threaten such a bombardment force and Slade did not consider the Japanese would risk a bombardment. Consequently, Singapore would be safe.

Hong Kong required separate consideration. Although the China Squadron of cruisers would be too weak to attack the Japanese battle fleet, it could, if left alone by the Japanese, operate from Hong Kong against Japanese convoys, scout to provide warning of Japanese sorties and force the Japanese to divert more ships to either mask or invade Hong Kong on their drive south. Any surface forces at Hong Kong could admittedly be masked by superior forces from the Pescadores Islands, although, if forced by their presence to move the lines of communication east of the Philippines and Borneo, Japanese ships would have to pass through the narrow Macassar Straits, where they would be vulnerable to attack by submarines and destroyers. In this case, Hong Kong would have served its purpose by forcing them off the direct route. Richmond, however, harboured doubts. He believed that the capture of Hong Kong would be the most likely aim of any Japanese aggression, and that its recapture would seriously delay the implementation of the British plans. He felt that the memorandum ignored this, presuming instead that the colony would be lost and that

> the strategical situation 'would not justify the use of naval forces in the defence of Hong Kong'. In my opinion the strategical situation which would result from the loss of Hong Kong does not justify withholding naval forces, if by their action they could contribute towards preventing its loss. If an addition of submarines to the squadron would enable some to be used in its defence, the claims of any other Station for their use must be very strong to override the claims of Hong Kong.[21]

Richmond felt that if Hong Kong was lost, the Japanese would make it impregnable in the shortest possible time, and its recovery by the British would only be possible by investment and a lengthy siege. The time it would take to bring about surrender would divert the fleet from its main aim: the neutralization of the Japanese fleet and the surrender of Japan. Once again, the viability of Hong Kong as a naval base became an important consideration.

Naval forces had to play a significant role in the defence of the colony, despite the long-established policy of not tying naval forces to local defence of bases.[22] If the other defences could not be improved and if the army would not strengthen the garrison, then the submarines of the China Squadron could be used to defend Hong Kong, forcing the Japanese to land some distance away from the colony, delaying operations and making the relief of the colony by the Main Fleet more likely. Cruisers could act against lines of communications and search out the Japanese battle fleet, operating from Hong Kong against Japanese sea lines of communication for as long as possible. The Main Fleet could then reach Hong Kong and act against Japanese trade more easily.

Slade revealed that the planning for Phases II and III was still at an early stage when he suggested that the East Indies Squadron of five or six light cruisers would proceed immediately on the outbreak of war to watch the possible lines of approach to Singapore, their place being taken by light cruisers pushed forward from the Mediterranean Fleet, whilst the China Squadron, another four or five cruisers, along with destroyers and submarines, operated from Hong Kong against the Japanese lines of communications. As stated previously, Richmond believed that this force would be better employed in providing a protective screen for troop convoys from India to Singapore after a declaration of war but before the Japanese acted. As both of these roles were also suggested by Slade, Richmond questioned what exactly his role would be. At one point, Slade had written that on the outbreak of war the Commander-in-Chief, East Indies, should 'immediately proceed to Sandaken or some other place in British North Borneo to watch the possible lines of approach of the Japanese either east or west of the Island',[23] and later in the same letter wrote that,

> As I see it, the duty of the CinC China is to find and occupy the Japanese line of communication while the CinC East Indies is mainly occupied at the outbreak of war in pushing forward the reinforcements [from India] as fast as possible and organizing the lines of communication both for trade and for war purposes.[24]

Slade may have been writing with the idea of the Commander-in-Chief ashore in overall control, but this was too vague for Richmond, whose handwritten comment on Slade's letter was simple and direct: 'But he isn't. See back. He is off Sandaken.'[25] In Richmond's mind,

Slade's proposals were unclear and took no account of the possibility of Japanese battlecruisers and armed merchant cruisers penetrating the Bay of Bengal, threatening British lines of communication and delaying the reinforcement to Singapore, a possibility that Richmond thought would need to be guarded against. Although it was probably unlikely that the Japanese would risk capital ships some 4,000 miles away from their home waters, if they did, their presence off the coast of India would have a disproportionate effect on the movement of trade, and on nationalist movements, not to mention the effects the sinking of troop ships laden with Indian soldiers would have. It was the prevention of this sort of move that Richmond saw as a crucial strategic consideration, more important than seeking a decisive battle:

> The political situation in India is none too stable as it is, and a great deal of the native unrest is said to date from the successes of Japan – the protagonist of the Orient, against the western power, Russia ... In the event of war direct action against the British will have far more effect in stirring up sympathetic movements in India than what might almost be called indirect action in attacking Australasia. That Japan rather allows the idea to be held that she has designs on Australia, while she is absolutely silent with regard to India, rather strengthens this view when we consider the careful, pro-Oriental, propaganda that is reported as being conducted from Tokyo.[26]

Japan was an island power, and it seemed certain to Slade and Richmond that it intended to push the nationalist movements to their utmost, and thus, threaten Britain in India, Burma, Borneo and Malaya. Slade hoped that the establishment of the Singapore base would stop Japanese aggressive expansionism, remove the threat of Japanese inspired anti-British disturbances in India and leave Japan to consolidate its position in China. Richmond remained to be convinced.

Richmond then went on to propose his own version of War Memorandum (Eastern), similar to the lecture he delivered whilst at Greenwich, and centred on Japan's Chinese and regional aspirations. He presumed that there would only be military action if Japan attempted political or economic control over China to the exclusion of other countries, Britain intervening only to force Japan to withdraw and stopping Japanese trade with China, and not attempting the economic exhaustion of Japan. The aim would be to convince Japan that the aggressive expansion of its influence was unprofitable and impracticable:

So long as she persists in her attitude she loses the trade with China she has hoped to increase: and if she cannot dislodge us from our controlling position she must abandon the policy that has brought about the war. We have, in fact, a strictly limited object, at any rate, in the initial phases: beyond that we cannot see.[27]

Occupying parts of China would mean that the Japanese would have to convoy troops there, which could only be maintained by sea, and these supply lines would become vulnerable to submarines operating from Hong Kong. To be successful, the Japanese must therefore take measures to render a British fleet unable to interrupt the lines of communication between Japan and China, whilst the objective of the British forces would be to control these communications between Japan and China. Richmond pointed out:

This cannot be achieved until several conditions have been fulfilled; the end can only be reached by stages. The objects of the intermediate stages are:–

(1) To hold Singapore and Hong Kong.
(2) To bring a fleet superior to the Japanese to the East.
(3) To establish the Base the fleet needs for its operations.
(4) To destroy or disable the Japanese fleet.

The ultimate British object would be unattainable if the British fleet were disabled... The British fleet will be disabled if it cannot bring force in sufficient quantity, when and where it is needed, successfully to oppose any movement of the Japanese fighting forces.[28]

So, whilst the ultimate strategic aim of the British would remain the destruction or neutralization of the Imperial Japanese Navy's battle fleet, the means of achieving this was slightly different. Until the arrival of the Main Fleet, British forces would be on the defensive, which might allow the Japanese the opportunity to seize Hong Kong, or act against Singapore, either one of which would give the Japanese a strategic advantage.

Despite all of the difficulties, Singapore would be the logical one to attack in force as its capture would oblige Hong Kong to surrender and would eliminate British trade in the China Sea whilst simultaneously securing Japanese trade. Moreover, once captured, it could provide a starting point for the occupation of Malaya and British Borneo and for attacks on the Fremantle area of Australia. Japanese battle cruisers could also safely attack trade in the Indian Ocean and threaten trade between Australasia and Britain. Most importantly, it would deny the British Main Fleet a base, and its recapture would become a long, costly and difficult task. It was therefore of vital importance that Singapore was as well defended and adequately garrisoned as possible to make such an attack a long and costly operation, impossible to undertake before the arrival of the Main Fleet from the Mediterranean.

But as Singapore was far from other Japanese bases, such an attack would not be launched unless the Japanese could confidently hope to destroy or disable the British fleet before it reached Singapore. Otherwise, the Japanese would probably revert to their traditional strategy and wait for the British to advance northwards and establish forward bases, when it could use its own strength to the greatest extent:

> Japan is unlikely to undertake action, with a view to disable the British fleet by battle, except under the conditions that are favourable to themselves, that is, in waters adjacent to Japan. If, however, the British force dispatched should have no decisive material superiority over the Japanese fleet, and if that superiority could be neutralized by engaging the fleet when short of fuel, weakened by sporadic flotilla attack, and in waters which confer tactical conditions favourable to the Japanese, the loss of the advantages of fighting in home waters might be preferred to the disadvantage of fighting a reinforced fleet later on, although under more favourable conditions to the Japanese.[29]

This was an argument against leaving any battleships in European waters, making it too risky for the Japanese to engage the British battle fleet anywhere in the region off Singapore. Using the lessons of history, Richmond recalled that

> To await the enemy at home in readiness and well supplied was advocated by Admiral Togo in the Tsushima campaign ... If the British fleet were inferior, or barely superior, it is possible, but not probable, that the Japanese would attempt battle at a distance from home waters. If the British had a marked superiority, it is most probable that the Japanese fleet would await battle in home waters.[30]

The danger to Singapore could only be countered by the early arrival of a relief fleet, within a month of the Japanese plan being set in motion, whilst the provision of adequate naval and military defences at Hong Kong and the provision of air units and a military force at both Hong Kong and Singapore to act in operations would help deter any other Japanese attacks, forcing the Japanese into other strategies, such as raids against British merchant shipping.

If, as was seen to be more likely, Singapore was not captured, the Japanese were expected to adopt cruiser warfare for exactly the same reason as the British, to force them onto the defensive and to make an unfavourable concentration with its battle fleet to screen the trade protection cruisers. This would allow the Japanese the opportunity of capturing Hong Kong, Borneo and Malaya, reducing the overall strength of the British battle fleet by sinking and damaging vessels and, at the right moment, engaging in a battle fleet action. If Japan failed in its attacks on British territories, it would have failed in the objective for which it went to war – regional superiority – and must then either make peace, or try to wear down the British will to fight through a prolonged war.

Rear Admiral Egerton, the Director of Plans, examined Richmond's criticisms in June 1925, and wrote to him in October. He agreed that War Memorandum (Eastern) needed further study by all three services, especially when Phase III was considered. However, up to that point, only Phase I had been considered in detail, and this was a phase where naval movements predominated. As for the concentration of the battle fleet, he pointed out that the proposed fleet was superior in gun calibre, not necessarily in guns. Richmond based his calculations on the five *Queen Elizabeth*-class, the five *R* class and the four battle cruisers opposing the Japanese battle fleet of six battleships and four battle cruisers. In total, the British battleships mounted 80 guns to the Japanese's 64, and the British battle cruisers 28 guns to Japan's 32, a total of 108 British heavy guns to 96 Japanese. If, as Richmond suggested, the battle cruiser *Tiger* was still working up after being recommissioned, and up to two battleships were in dockyard hands for repair and refit, then the British total of guns would be around 84, clearly inferior to the Japanese in number, if not in calibre.[31] The margin of superiority was small, however, and Richmond felt that this was too narrow a margin to begin a war with Japan, especially if it was proposed that some of Britain's battleships stay in European waters.

Egerton did not agree with this analysis, although he did agree that Phase I should not end at Singapore, and that Phase II might begin at either Singapore or Hong Kong. But he felt that this should be a decision left to the Commander-in-Chief on the spot, rather than be written down and regarded as the only course of action. However, the security of Singapore was the primary objective, and must remain so, especially as Japan's intentions could not be known.

Shortly afterwards, Egerton informed Richmond that the agenda for the Singapore conference was approved and that Phase III of the plan would be considered:

The actual memorandum itself is, I fully realize, a sketchy one but it is the first attempt from here to show that the whole business must be thrashed out from the economic aspect. Meanwhile we are now getting on with Phase I and the passage of the Fleet will soon be an organized affair with the result that interest will be focused on certain aspects of it for which no adequate preparations have been made.[32]

The plans had been discussed by the regional Commanders-in-Chief at Penang, and in December the Staff College at Camberley set out to test the scenario where the Japanese had captured Singapore before the arrival of the Main Fleet, which had now to set about its recapture. Egerton informed Richmond that

we hope to illustrate that which is fairly obvious, namely that the immense operation of its recapture will never have to be undertaken. ... Of course it is a scheme visualizing what might happen in 1925, 26 or 27, but not at a later date when we hope to be able to meet force with Japan.[33]

Richmond believed that the exercise at Salsette Island had already shown that the difficulties of launching an invasion attempt on Singapore were so great as to make it unlikely and that, instead, the Admiralty should have a range of objectives developed to wear Japan down, rather than rely on the prospect of the Japanese wanting to fight a decisive naval battle.

As a further part of the planning process for a war in the Far East, in March 1925 the Commander-in-Chief of the Mediterranean Fleet wrote to the Admiralty about the passage of his fleet to Singapore.[34] At any one time, he expected to have one battleship and one cruiser either refitting in home waters, or recommissioning, and one battleship, one cruiser and possibly one destroyer flotilla refitting at Malta, as well as a destroyer flotilla and a depot ship in the Eastern Mediterranean. So not all of his fleet might be immediately available.

In the event of relations between Japan and Britain becoming strained, he expected his fleet to concentrate at Malta and take on supplies, fuel and aircraft, whilst the flotilla in the Eastern Mediterranean, along with the Red Sea sloops, would secure the Suez Canal from attack. On receipt of the warning telegram, the fleet would sail to the Suez Canal and transit, at 7 knots. At this speed it would take 30 hours for a battle squadron and 20 hours for a cruiser squadron or destroyer flotilla to pass through the Suez Canal. Once through, they would proceed to bases M and T, and from there to Singapore, at a speed of 16 knots. Obviously, during this stage, movements would be dependent on the intelligence of Japanese actions and movements. If the Japanese had declined to attack Singapore the pace of British operations would be far different. If the Japanese battle cruisers were raiding in the Indian Ocean, the valuable fleet train of tankers and colliers would need to be safely convoyed and protected. A total of 42,880 tons of oil and 15,300 tons of coal, carried in six tankers and four colliers, would be needed at M and T, with more at Trincomalee. Even in ideal conditions, it would be some considerable time before the fleet could reach Singapore.

More ominously, another review of the Mediterranean Fleet's readiness for war, conducted in 1927, concluded that the individual ships were ready, but the fleet as a whole was not.[35] It was unable to deploy instantly, could not seize an advanced base until some weeks or months after war had started, and had no net layers, indicator nets or mines on station. In addition, the mobile guns and searchlights for the Mobile Naval Base Defence Organization were lacking, and ammunition, stores, *X* lighters, monitors and motor torpedo boats were all needed, along with ASDIC sets and suitable vessels for anti-submarine work. This inability to deploy speedily would have had serious implications had a war with Japan broken out.

War Memorandum (Eastern) was updated in August 1927. Although the plan remained largely the same, it was recognized that

> The maintenance of heavy ships in the East will not, however, be possible until the floating dock at Singapore is in place, even then there will be no crane capable of lifting big guns and the storage facilities will be so small that the amount of spare gear available must be very limited whilst there will be no armament depot for the storage of cordite, shell etc. Consequently, the dispatch of Battle cruisers to the East could only be justified by an adverse political situation so long as maintenance facilities are so limited.[36]

This meant that in time of war the Pacific Fleet would be based around cruisers and submarines only, and on these ships would fall the burden of trying to defend Hong Kong and Singapore, as well as protect the lines of communication until the Main Fleet arrived. Two battleship squadrons, two cruiser squadrons, two aircraft carriers and four destroyer flotillas would be proceeding from the Mediterranean, one battleship squadron, one battle cruiser squadron, one cruiser squadron, two aircraft carriers and two destroyer flotillas, in addition to two submarine flotillas and the minelayer *Adventure* would sail from the Atlantic, joining ships already on station.[37]

The strain on dockyard facilities would be great, as apart from the Main Fleet, these facilities would be needed to maintain an additional 16 cruisers, 46 armed merchant cruisers and many destroyers, submarines and auxiliary craft. All of the docking facilities at Hong Kong, Keppel Harbour, Calcutta, Bombay and even Sydney and Melbourne would be needed.[38]

The next major revision, the 1931 version of War Memorandum (Eastern), addressed some of Richmond's concerns and placed slightly less emphasis on the decisive battle as an end in itself, and more on the importance of economic pressure. The Naval Staff had gathered information in 1929 and concluded that the Japanese industrial base would be unable to withstand a dislocation of trade in the event of a war. But it still kept the decisive battle as the way of defeating Japan.[39] The Plans Division therefore also determined that it would be impossible to inflict any damage on Japan's trade until its battle fleet had been neutralized and that this goal would be best achieved by forcing the Japanese to run some sort of risk with its battle fleet, possibly by joint operations against Japanese forces in China, attacks on their sea lines of communications, applications of economic pressure, a direct attack on the Japanese fleet in harbour or a direct attack on Japanese forces occupying British territory.[40] Thus,

> Any successes against these [Formosa and the Pescadores Islands] is likely to bring the Japanese fleet to the southward whether it be to relieve the pressure by fighting a fleet action or to cover the arrival of military reinforcements. If the Japanese refuse to expose their Main Fleet to the risks of a fleet action we can proceed with our progressive campaign to the northward until they are forced to accept action.[41]

The cooperation of the other armed services was still presumed to be readily available. Singapore was still judged to be secure, provided the defences were developed, and although a raid was still possible, distance, and the risks involved, remained one of the best defences against any full-scale invasion, which was judged likely to fail.

Many of these points had been considered earlier, raising the question as to whether, in developing a strategy to fight the Japanese, the Royal Navy was using the prospect of fighting the Imperial Japanese fleet as both a rationale for a large battle fleet and a measure of its own ability to tackle any unforeseen event that might arise, including, of course, just such a war. Strategy that was not continually being questioned or which could not be carried out by the available forces would inevitably break down under the stresses and strains of war; this is where the importance of the preparation strategy lay. War Memorandum (Eastern) was one such example of the preparations.

Fuel provision remained a problem, even with the output of the refineries at Abadan and Rangoon, and even with buying up the entire Netherlands East Indies output. It was calculated that over 180 oilers would be needed, and because battleships could not transit the Suez Canal fully loaded, there would need to be adequate fuel supplies at Port Said, Aden, Colombo and Rangoon. Tankers would therefore be needed to supply these sites, and they would need escorts. Both tankers and oilers would need fuel for themselves.[42] If the coal-burning *Iron Dukes* and the battle cruiser *Tiger* were sent, they would be dependent on coal carried in colliers with an average speed of 9 knots.[43]

Supplies of approximately 1,250,000 tons of oil were envisaged for Singapore by 1933 which would give ample mobility up to 1,500 miles from Singapore. But if the battleships consumed 25,000 tons of oil per month, three battle cruisers 15,000 tons, 15 cruisers 30,000 tons, 36 destroyers 36,000 tons and 21 submarines 2,100 tons, this would present a large fuelling problem, requiring nearly 110,000 tons of oil fuel per month, which would need replenishing from escorted convoys every seventh day. An additional six small tankers would be required for an advanced base, a further five oilers as base oilers and five oilers at the advanced base for fleet work. In total this would add up to 20 oilers for the advanced base alone. Without adequate stockpiles at Singapore, and on the route to it, which had already been acknowledged as a priority for development, operations by a battle fleet would be impossible. But the number of tankers which would be needed could not do other than have a serious effect on Britain's imports of oil and would be another formidable undertaking.

Despite all of the difficulties, War Memorandum (Eastern) helped the Royal Navy define itself in the 1920s, in terms of who its most likely enemy was, the difficulties which needed to be overcome to meet that enemy in combat successfully, the types of vessel required in the fleet, and the tactics necessary to fight them and win. With Japan as the only conceivable enemy during the 1920s, it helped to develop a strategic policy for the Navy and strengthened the claims that Singapore had to be developed into a fully functioning naval base. The defeat of the Japanese armed forces would be a decisive operation and the strategic planning needed to be more than just a plan for the deployment of ships. The Plans Division initially assumed that Japan would be vulnerable to economic blockade, just as Britain would be, both countries being islands. They also believed that by disrupting this trade with cruisers, the Japanese battle fleet would be compelled to put to sea and fight for maritime control.[44]

Whether such an undertaking was actually possible at the time of planning, 1924, is another matter, although if Britain and Japan had gone to war in the 1920s, then the British undoubtedly would have implemented the plan, as best they could, with whatever ships and resources were available.⁴⁵ At the heart of the planning, however, lay an assumption of British superiority, resting not only on the Royal Navy's numerical superiority, guaranteed by the Washington Treaty, but on the belief that the training was better, that the ships were better and that the men were better, so that even a numerically equal British fleet would always be superior to a Japanese one in battle. The lessons of economic warfare in the First World War were still largely ignored, even though the Naval Staff believed that Japanese industry was ill-equipped to withstand the ravages of a long trade war, and that Japan lacked most of the necessary raw materials to survive a blockade. Trade warfare was still secondary to a battle fleet action because it was believed that the effectiveness of any British blockade would turn on how completely links between Japan, China and the United States could be cut, and on the fact that this disruption of Japanese trade, especially in the China Sea, could not be carried out until the Japanese battle fleet was neutralized.

Although Richmond made some lengthy criticisms of the Admiralty plan, his ideas did not actually differ greatly from those of the Admiralty. Both thought mainly in terms of a naval campaign, with superior British forces containing or destroying the Japanese, and both saw Hong Kong and Singapore as essential for a successful war against the Japanese. As Commander-in-Chief, East Indies, Richmond recognized the importance of involving the army and RAF in the planning, if the strategy was to succeed in defeating Japan. He believed that the naval forces at his disposal would be better used in escorting reinforcements to Singapore from India, rather than being wasted on attacks on the Japanese invasion force, screened as it would be by the Japanese battle fleet. The main differences between him and the Plans Division concerned what would happen once the fleet was at Singapore.

Ultimately, sea power was dependent not on the size of the fleet but on the strategic position. The battle fleet was the tactical factor, geographical position the strategical. The connection between the two was the strategic will, operating through the strategic operations plan and guiding the fleet to a strategic position. War Memorandum (Eastern) provided the strategical will and, to a large extent, the *raison d'être* for the Royal Navy. By the end of the decade, it had developed into a plan that attempted to provide a set of likely possible causes for a war, and how and where the Royal Navy was going to operate in response. Having the strategy also helped to justify the need for the Singapore naval base, a need endorsed by Australia and New Zealand, and it also provided a context for the battle fleet and related tactical developments.

ADMIRAL RICHMOND AND WAR MEMORANDUM (EASTERN)

NOTES

1. PRO, ADM 116/3125, 'War Memorandum (Eastern); General Remarks, Part III: Suggested Line on Which War Memorandum (Eastern) should be prepared', pp. 4–5.
2. Richmond Papers, RIC 10/5b, 'Strategy 6: Cooperation, 23 February 1921'.
3. Richmond Papers, RIC 10/5a, 'Introduction to Strategy Lectures, Spring 1921, 15 March 1921'.
4. Richmond Papers, RIC 10/5b, 'Strategy 3; Object and Function of the Fleet, 18 March 1921' and 'Strategy 4: Influence of Strategy on Materials, 21 March 1921'.
5. Fiske, *The Navy as a Fighting Machine*, p. 269.
6. Richmond Papers, RIC 11/1, 'Supply of Fleets and Bases. Captain N.F. Lawrence, RN, 22 March 1921'.
7. Bell, 'How Are We Going to Make War?', p. 12.
8. PRO, ADM 116/3125, 'War Memorandum (Eastern): General Remarks, Part I', p. 5, Part II, p. 8.
9. PRO, ADM 116/3125, 'Remarks on Paragraphs in Detail', Part II, p. 11.
10. Ibid., p. 14.
11. Ibid., pp. 16–17.
12. NHB, W.J.R. Gardner to author, 26 August 1999.
13. PRO, ADM 203/84, 'Combined Naval and Military Exercise Carried Out on Salsette Island, Bombay, December 1924', p. 2.
14. Richmond Papers, RIC 7/3e, 'Letter from Egerton to Richmond, June 1925'.
15. PRO, ADM 203/84, copy of a Report from Commander-in-Chief, East Indies, p. 3.
16. Ibid., p. 9. Also see R. Harding, 'The Royal Navy and Amphibious Operations, 1919–1939', 'Adapting to Change: The Royal Navy and the Maritime Industries, 1815–1990', p. 9, Conference organized jointly by National Maritime Museum and University of Westminster, 31 October 1998.
17. Ibid., p. 5.
18. NHB, *Godfrey Papers*, 'Mobility', Staff College Lectures, 1924–25, Royal Navy Staff College, Greenwich, 29 October 1924.
19. PRO, ADM 116/3125, 'Remarks on Paragraphs in Detail, Part II', p. 5.
20. NMM, Richmond Papers, RIC 7/3e, 'Slade to Richmond, 22 March 1924'.
21. PRO, ADM 116/3125, 'Remarks on Paragraphs in Detail, Part 1', p. 3.
22. Ibid., 'Part II', p. 5.
23. NMM, Richmond Papers, RIC 7/3e.
24. Ibid.
25. Ibid.
26. PRO, ADM 116/3125, 'War Memorandum (Eastern): General Remarks, Part III: Suggested Line on Which War Memorandum (Eastern) should be prepared', p. 1.
27. Ibid., pp. 4–5.
28. Ibid., p. 6.
29. Ibid., pp. 7–7a.
30. Ibid., pp. 100–1.
31. NMM, Richmond Papers, RIC 7/3e, 'Egerton to Richmond, 21 October 1924'. The British ships mounted 15-inch guns as well as 13.5-inch on the older *Tiger* and *Iron Duke* class, as well as 16-inch guns on the *Nelson* class, once they were commissioned. The Japanese ships mounted 14-inch guns, with the exception of the two *Nagato*-class ships, which mounted 16-inch guns.

32. Ibid., 'Egerton to Richmond, 18 November 1924'.
33. Ibid.
34. PRO, ADM 116/3125, 'War Memorandum (Eastern): Appendix C; Passage of Mediterranean Fleet to the Far East'.
35. PRO, ADM 116/3134, 'Mediterranean Fleet: Preparations and Readiness for War, 6 January 1927'.
36. PRO, ADM 116/3125, 'War Memorandum (Eastern): Corrigenda No. 2 to War Memorandum (Eastern), 10 August 1929', p. 189.
37. PRO, ADM 116/3125, 'Appendix C', p. 19.
38. Ibid.
39. Bell, 'How Are We Going to Make War?', pp. 131–5.
40. Ibid.
41. Bell, *The Royal Navy*, p. 75.
42. NHB, *Godfrey Papers*, 'Supply I and II, 1924–25 Session', p. 1.
43. Ibid., p. 22. Although usually in Reserve, these ships were regularly recommissioned to replace modern ships undergoing refit. They certainly could have formed part of a battle fleet until 1931, by which time they were due for disposal as a part of the Washington Agreement, with the exception of the *Iron Duke*, which was to become a gunnery training ship.
44. C. Bell, 'The Royal Navy, War Planning and Intelligence between the Wars', pp. 3–4, in J. Siegel and P. Jackson (eds), *Intelligence and the International System* (Praeger, New York, forthcoming).
45. In 1982, facing an unexpected act of aggression and with no specific, pre-prepared war plan, the Royal Navy formed, at short notice, a Task Force that sailed and was successful in recapturing the Falkland Islands.

4

Developing the Far Eastern Strategy: War Memorandum (Eastern) and Changing Circumstances, 1931–41

Sending the battle fleet to Singapore was only the first stage in projecting British power against Japan, and although the first versions of the war plan focused primarily on a decisive fleet action, changing circumstances both at home and overseas forced the Plans Division to recast their strategy as increasingly they acknowledged that there would be a prolonged period of attrition as Japan's economy was slowly squeezed, and Japan was forced to seek terms. Richmond's critique of War Memorandum (Eastern) had stressed this, and the 1931 memorandum began to reflect this. But such a war was only possible if the British had forward bases to operate from and there were sufficient ships, especially cruisers, available to implement it. The Admiralty, therefore, developed several versions of War Memorandum (Eastern) between 1919 and 1939 which, of necessity, reflected the changing international naval and strategic situation, as well as the increasing domestic constraints caused partly by the financial situation with which Britain found itself faced.

The Admiralty plans stressed the key role Singapore could have in the event of a war between Britain and Japan, and also between the United States and Japan. It was thought unlikely that either side could bring sufficient military or economic pressure to force a conclusion, and also as it was thought improbable that the United States would risk sending its fleet to the Philippines, a distance of nearly 7,000 miles, until such times as an advanced naval base was established there. Moreover, with the Washington Agreements affording no possibility of fortifying Guam or the Philippines, Britain could present itself as an arbiter in any war between these two powers, if not an outright ally, allowing US ships the use of the base, an option that became increasingly attractive to the British as the 1930s progressed. But as long as British interests in the Far East were not directly jeopardized, it was advantageous for the British Empire not to be involved in such a conflict. However if Japan seized the Philippines, bringing it nearer to British possessions in Borneo, Malaya, Australasia and New Guinea, and nearer to India, this might present such a threat. Coupled with Britain's growing inability to deploy a sufficiently strong battle fleet to the Far East, whilst keeping sufficient ships in home and Mediterranean waters to balance the growing Italian and German threats, this made basing units of the US Pacific Fleet a more attractive alternative for Britain's strategic planners in the late 1930s.

And as the 1930s progressed, the political focus swung to Europe and the Mediterranean, highlighting the Royal Navy's difficulties in deploying battle fleets to more than one area of operations, and calling for a reappraisal of strategic priorities and plans. The Admiralty, initially, was not prepared to act on the deterrent principal of sending only a small squadron to Singapore, and even the proposed 'Peace Fleet' of 1923 was seen more as a vanguard than a final fleet. The idea of sending smaller battle fleets, and even a flying squadron, was one forced on the Admiralty by necessity, along with increased efforts to persuade the Americans to deploy their battleships to Singapore.

The driving force behind the Far Eastern war planning during the 1920s had been the First Sea Lord, Admiral Sir David Beatty, convinced, as he was, of the dangers posed by Japan. As First Sea Lords after Beatty, both Admiral Sir Charles Madden and Admiral Sir Frederick Field moved away from an explicit Far Eastern strategy towards general principles of sea warfare for the Royal Navy, principles which reflected both the smaller fleet and the increased demands they were dealing with. Madden, First Sea Lord between 1927 and 1930, for example, was unable to stand up to Ramsay MacDonald's pressure over cruisers at the time of the London Naval Conference, despite the protestations of Field, his DCNS and Chatfield, the Controller of the Navy, who had urged him to press for 70 cruisers, including heavy cruisers armed with 8-inch guns, to counter foreign construction. But although Madden bowed to political pressure, he insisted that the cruisers be new ships, relying on speed and volume of fire rather than weight of shells, and with a large radius of action and the ability to operate in all parts of the world, including the Far East.[1]

Field, his successor, in post between 1930 and 1933, continued giving the government a clear statement of Britain's naval requirements. Field advocated keeping a strong fleet in home waters as a deterrent, not only to counter possible aggression by Italy and Germany, but also to guard against any possible Japanese aggression in the Pacific. So long as Britain had only one enemy to deal with, such a fleet could be sent out east as and when the circumstances dictated. In such circumstances, the strength of the Royal Navy still seemed adequate for such a task and the Admiralty was confident as to the outcome. But in order to be able to operate such a fleet the Royal Navy had to be certain of having secure bases at Singapore and Hong Kong.

However, Admiralty fears over the uncertainty of Britain's defence preparations in the Far East seemed to be borne out when, on 5 January 1927, the British Concession at Hankow was overrun by a mob including Cantonese troops, and it seemed only a matter of time before Shanghai was also overrun. A Field Force, comprising three infantry brigades, was sent immediately to Hankow to defend British interests. There were no tanks, armoured cars or aircraft available to be attached to it and Lord Milne, the Chief of the Imperial General Staff, was led to express his concerns over the defence of British possessions such as Hong Kong and, to a lesser extent, Singapore.[2] The very difficulties the army faced strengthened the Royal Navy's case that a strong battle fleet was the most suitable defence for Britain's Far Eastern possessions and war in the Far East was portrayed as primarily a naval affair, with the army and RAF playing a minor role and providing the necessary forces for the advance northwards.

However, by 1933, Chatfield, who followed Field as First Sea Lord, was having to inform the Chiefs of Staff that the European situation was changing and that whilst the Far East was still a potential danger zone, the rise of fascist regimes in Italy and Germany was contributing to a deteriorating European situation.[3] At this time, Germany's navy was not a major factor in naval strategic planning, although there were concerns about the qualities of the *Panzerschiff*, 'pocket battleships', as commerce raiders, and when the Defence Requirements Committee was set up at the end of 1933, in a decision which marked a fundamental, strategic shift of thinking, Germany was identified as the ultimate, potential enemy, even though Japan seemed to present the more obvious and immediate threat.[4]

Intelligence, or increasingly the lack of it, was now becoming an important consideration for the Admiralty's war planners. They needed to know what Japanese strength was in order to assess the immediacy of the possible threat and it was increasingly difficult for British naval attachés to make accurate assessments of the Imperial Japanese Navy's men and ships. In 1933, for example, the Director of the Naval Intelligence Division wrote to Captain Ross, the Naval Attaché at the Tokyo Embassy:

Our Intelligence Service has found it increasingly difficult to get any information concerning their Armed services. Our Confidential Book on Japan is some thirty years out of date. We know little about their warships – they could build a new battleship or aircraft carrier without our knowing ... I want to know how they think, how they intend to operate their fleets at sea ... Report on their shipbuilding programmes. What is their fuel situation? Also study what their Navy is doing in aviation.[5]

At the same time as the DNI was bemoaning the lack of accurate intelligence, the Admiralty apparently received secret information regarding a Japanese scheme for attacking Singapore. According to a memorandum prepared by Chatfield in February 1933 for the Chiefs of Staff Committee, there was, reputedly, and had been since the Shanghai Crisis of 1932, a Japanese division, with light artillery, and 18 transports ready to sail from a Japanese base in the Pescadores Islands, 8–10 days' sailing away, via the Pelew Islands, to the south of the Philippines, to the north of Borneo and thence to Singapore.[6] If this report was true, the Far Eastern situation could quickly deteriorate into war.

The Committee therefore recommended completing the defences of Singapore within the following three years to safeguard any possible Far Eastern strategy and then starting preparations for a European war. This seemed to be a sensible solution whilst the government waited to see how the situation in Europe developed, and to determine how the British should respond to them. After all, if Germany was going to rise again as a European power, and Britain's emphasis was on developing the defences in the Far East, then Britain itself could be at risk. Even so, there was the confident belief that in the event of a war with Germany, the RAF and the French Army would be able to take the initiative against Germany without the assistance of a British force and that full and immediate mobilization of Britain's army within six months of a war being declared was adequate.

Nevertheless, the Cabinet decided to abandon the Ten Year Rule, and defence expenditure in the future was to be based on the priorities of the Far East, Europe and India. Neville Chamberlain, the Chancellor of the Exchequer, did not believe that Britain could afford to fight the two-front war suggested by such a priority, and that the best and most cost-effective deterrent was a strong, defensive air force. To pay for the build-up of the RAF it would obviously be necessary to cut back on naval preparations. As a contribution to this, Chatfield successfully advocated the signing of the Anglo-German Naval Agreement, which tied Germany to having a fleet which was

35 per cent the size of the Royal Navy, allowing the latter to face a German threat if it appeared, as well as meeting its Far Eastern commitments. Indeed, Dudley Pound, a future First Sea Lord, wrote of the Agreement that it removed a stumbling block – having to keep sufficient forces in home waters to watch the growing strength of the German Navy – to Britain's naval, strategic planning.

This was at a cost. The agreement between the two powers relieved the pressure on the British for a while, but the agreement finally marked the end of the Versailles settlement as a means, however imperfect, of trying to restrain German expansionism. Even with the recent naval agreement, when Germany's naval plans, and those of the French and Italians were completed, and until its own new ships were completed, the Royal Navy would have no battleships to equal their new ships in tonnage, speed and gun power, and would need new, modern vessels as soon as it could possibly build them. The threat of German naval action against Britain in the Atlantic Ocean was much more immediate than a war in the Far East could be.[7]

However, it was Italy, Europe's first fascist power, and not Japan, that became the most serious strategic consideration when, during the 1935 Abyssinian Crisis, the British felt unable to oppose the Italians, fearing that Italy was in a position to cut essential lines of trade and communication across the Mediterranean, whilst the Royal Navy did not have a secure, defended base at either Malta or Alexandria, nor the ability to pass vessels through the Mediterranean Sea. (Paradoxically, at precisely the same time, the Regia Marina felt unable to tackle the Royal Navy in the Mediterranean in the event of a war.[8]) As Chatfield's overriding concern was to avoid any conflicts whilst the country, and especially the Royal Navy, was building up and rearming, no aggressive action could be taken. Quite simply, Britain was caught at a vulnerable time, with its navy in the throes of modernization. And with estimates showing that the Imperial Japanese Navy would be at its strongest by 1945, the Royal Navy would have been incapable of fighting a war against both Italy and the Japanese, who were expected to seek to take advantage of a war between Britain and Italy by attacking in the Far East.

Nevertheless, from 1935 onwards the Naval Intelligence Division was constantly assessing the likelihood of a war with Italy, and whilst Pound, the Commander-in-Chief of the Mediterranean Fleet, was confident about his command's ability to deal with the Italian battle fleet, the Admiralty was also considering what the possible reactions of the Japanese would be in the event of a war, and how they would find a battle fleet to go to Singapore, without abandoning the

Mediterranean Sea if, as expected, the Japanese declared war whilst the British were embroiled in a war in the Mediterranean Sea.[9]

As the magnitude of this problem became more apparent, Far Eastern war plans continued to be recast to reflect the more unstable European situation. The defence and utilization of Hong Kong remained unresolved and, in 1931, 1933 and again in 1937, the Admiralty, unlike the army and RAF, believed that Hong Kong could function as a base for light forces, despite its poor defences, operating against Japanese trade. In 1933 the Admiralty had informed the Commander-in-Chief of the China Station that Hong Kong was a necessary base from which to operate to protect Britain's interests in China and that without it the Royal Navy could not operate in Japanese waters.

And despite the lack of enthusiasm for improving the defences of Hong Kong from the army and the RAF, the Admiralty planners clung to the vision of cruisers and destroyers, possibly accompanied by US ships, acting against Japanese lines of communication. A Chiefs of Staff review in 1938 rejected this assumption, however, as Japan's continued expansion into China made the successful defence of Hong Kong more and more unlikely, and they were only prepared to agree to its defence as a means of denying it to the Japanese. As Admiralty planners had to come to terms with the fact that it was becoming less and less likely that the Royal Navy could send a superior fleet to the Far East in time of war, and given that Japanese forces were now occupying much of mainland China, a war plan that included an advance northwards, from Hong Kong, was no longer realistic.[10]

As a consequence, in late 1937 a revised version of War Memorandum (Eastern) considered several options once the British battle fleet had arrived at Singapore, and it now included the possibility of a European situation making the strategy defensive. It was now assumed that Hong Kong could only be held if the defences were on a par with Singapore's and, instead, the evacuation of Hong Kong was now advised, as the dangers of air attack made it unlikely that it could be adequately defended, let alone used as an advanced fleet base.[11]

If Hong Kong fell into Japanese hands, the fleet was to concentrate on Singapore and hope that the Japanese would be drawn southwards by British attacks on Japanese possessions, such as Formosa and the Pescadores Islands, as well as on Japan's maritime trade, especially its oil tankers, within striking range of the fleet and fixed defences of Singapore, thus enabling the Royal Navy, supported by the RAF, to attack and seriously damage enough Japanese units to deny the Japanese control of the South China Seas. It would then be possible to send reinforcements and slowly commence the advance northwards, exerting more economic pressure on Japan, significantly *without* the need for widespread fleet actions. With Japanese vessels being captured off the coast of the United States and in European waters, Japan's trade would be cut off and its economy strangled, and, slowly, it would be forced to surrender. Even this level of planning seemed to be more and more unrealistic, taking no account of possible Japanese intentions and depending more on good fortune than strategic thought.

Of necessity, between 1935 and 1939, the focus of British naval strategy had inevitably swung away from the Far East and back to Europe and the Mediterranean Sea, as German and Italian naval expansion and the threats to Britain's lines of communication through the Mediterranean Sea and Suez Canal preoccupied the Admiralty. Two years after the Abyssinian Crisis, in 1937, the Naval Staff indeed pointed out that the Singapore strategy was now a hostage to fortune, and that given the political situation in Europe it was impossible to send anything other than light forces to the Far East in time of war, unless the political decision was made to abandon the Mediterranean Sea.

As if to emphasize the scope of the problem facing the British, on the same day that the USS *Panay* was bombed and sunk, on 12 December 1937, the gunboats *Ladybird* and *Bee* were fired on by the Japanese on the Yangtze and the *Scarab* and *Cricket* were bombed near Nanking. Both the United States' and British governments were incensed by these attacks, and other high-handed actions of the Japanese, including the mounting evidence that the Japanese were trying to squeeze them out of the China trade. For their part, the Japanese claimed that Britain and the United States were actively supporting the Chinese against them, rather than just trying to exercise their rights to trade as neutrals. But, in the view of the Chiefs of Staff, there was certainly no doubt that Britain was unable to face a war against Germany, Italy and Japan simultaneously.[12]

It was judged more likely that Japan would declare war when Britain was embroiled in a war with Germany and Italy, and existing naval forces in the Far East, primarily the cruisers of the China Squadron, were adequate to protect British interests, but nothing more. The Admiralty still aimed to establish a battle fleet at Singapore, although only within 70 days of a war breaking out, and the strength of the proposed fleet in 1937 was to be 12 battleships (the entire strength of the Royal Navy, minus *Valiant, Renown* and *Queen Elizabeth*, all of which were being modified) to oppose the nine Japanese. Based on Singapore, with at least two ships refitting at any one time, this would leave a fleet of ten available ships, fewer if the four battleships needed to watch the growing German naval strength were taken out. But this would not allow for any British battleships in the Mediterranean Sea to face the Italians. Only the French battle fleet was available to do this, and withdrawing Britain's naval presence in the Mediterranean was not seen as politically acceptable. It made more sense to remain in the Mediterranean and try to persuade the Americans to become more actively involved, with their fleet, in the Far East.

But, despite this, the Dominions were constantly reassured by the Chiefs of Staff and by Backhouse, when he was First Sea Lord, that a British battle fleet would be sent eastwards when needed. Backhouse intimated that, strategically, the Mediterranean was of secondary importance, and in November 1938 the Australian High Commissioner in London was given what he regarded as an unqualified assurance from Backhouse that in the event of a war with Japan seven capital ships would be sent eastwards. But, in March 1939, in reply to further Australian enquiries, the Prime Minister, Neville Chamberlain, more realistically telegraphed that the size of any fleet sent eastwards would be dependent on at what stage Japan entered the war, and losses to date.

The Japanese occupation of Hainan and the Spratley Islands in 1939, only 650 miles from Singapore, confirmed British fears of Japanese expansion and raised the question, given that it was impossible to send a fleet to Singapore, of how Britain's interests could be best represented. Economic reprisals may have led to war, the Concessions in China were at risk of Japanese attack and US support was vital before any action could be taken. Japan needed to be appeased as much as possible, but so did Italy and Germany, and the Admiralty's Far Eastern strategy was feasible only if there were no European distractions. Otherwise, the strategy of sending a fleet to the east in the event of Japan declaring war was not a workable option. Even if it were, Singapore's period before relief was now calculated to be 90 days, and in September 1939 this estimate was extended to six months.

By 1938, the Committee of Imperial Defence felt that even with rearmament and the 'New Standard' fleet in place, the Royal Navy would be hard put to maintain naval superiority over the combined fleets of Germany, Italy and Japan. And with the arrival of Sir Roger Backhouse to the post of First Sea Lord in 1938, British naval strategy underwent its most profound of shifts, firmly *away* from the Far East, as the Admiralty became increasingly pessimistic about being able to send a fleet to Singapore. In February 1939, during a meeting of the Strategical Appreciation Sub-Committee, which discussed the despatch of a fleet to the Far East, Cunningham, DCNS, pointed out that the Royal Navy needed six ships in home waters to guard the three *Deutchsland* pocket battleships, leaving three for the Mediterranean and six for the Far East. This was not an adequate fleet to face the ten Japanese battleships, although it would be somewhat better if these six were fast ships. However, Britain's battle cruisers were the only ships with the combination of speed and firepower to match the German ships, meaning that any force destined for Singapore would have had to be composed of the three modernized *Queen Elizabeth* class, supported by the slower *Nelson*, *Rodney* and one *R* class, perhaps later reinforced by more of the *R* class and the battle cruiser *Renown*. Even gathering this fleet together was not without difficulty, as the First Sea Lord pointed out:

> the process of disengaging the Fleet from the Mediterranean and home waters during war with Germany and Italy and the formation of a Far East Fleet, with its accompanying cruiser squadrons, destroyer flotillas and auxiliaries, would be an operation of magnitude and could not be carried out at a moment's notice.[13]

This was a far cry from the situation as envisaged in 1924. In another memo, 'The Despatch of a Fleet to the Far East', the DCNS pointed out that

> the Chiefs of Staff have repeatedly stated that, in their opinion, our present and potential naval strength is insufficient, and indeed is not designed to engage three naval powers simultaneously ... there are so many variable factors which cannot at present be assessed, that it is not possible to state definitely how soon after Japanese intervention a Fleet could be dispatched to the Far East. Neither is it possible to enumerate precisely the size of the Fleet that we could afford to send.[14]

The number of capital ships available was limited, and it appears from this that any realistic possibility of sending a battle fleet to Singapore was dead.

But, despite this, in a meeting of the Committee of Imperial Defence in April 1939, Japan was assumed to be a hostile neutral in the event of a European war, and if it declared war, it was agreed that a fleet would have to be sent eastwards. But, before deciding on its strength and composition, other operations, such as those against Italy in the Mediterranean, the necessity of offering support to allies, the course of Japanese operations and the possibility of US involvement would all have to be considered. Italy remained the key:

If Japan delayed entry into the war, it might be possible to establish a measure of control in the Mediterranean, sufficient to allow for the immediate release of British forces for the Far East. A similar result would also be obtained if serious losses had been inflicted on the German and Italian Navies.[15]

This would leave five or six battleships available for the Far East, still an inadequate force to meet the Japanese battle fleet of ten ships, supported by cruisers, destroyers, submarines and land- and carrier-based aircraft. British naval forces in the area were totally inadequate for the defence of Hong Kong and Singapore, and the available reinforcements inadequate to throw back the Japanese. The situation was indeed becoming dire, with the possibility of the small China Squadron being overwhelmed by a sudden Japanese attack.

All of this affected the overall strategic and operational planning. During his term as Commander-in-Chief, China Station, Admiral Frederick Dreyer was responsible for the gathering of information regarding suitable anchorages to the north of Singapore and the level of defence they would need if the Royal Navy was to use them in the relief of Hong Kong. But Dreyer determined that it was Hong Kong, and not Singapore, that held the key to success in the region. Singapore was *strategically* important, but was only vital if a battle fleet was able to operate against the Japanese, whereas Hong Kong would be a vital base from day one of any war, and especially if the Royal Navy was going to take an early offensive against Japanese vessels trading between China and Japan. In his analysis, he wrote that the British objective, securing their Far Eastern possessions, would be best achieved by using their naval power from both Hong Kong and Singapore. Both bases were necessary, and yet both still needed their shore defences building up. In the meantime, only naval forces could defend them.[16]

Dreyer, whilst engaged on gathering information about anchorages, had developed a less ambitious strategy, which he called Plan X, based on the concept of denying access by sea to the Japanese to the south and west of a line from Malaya to Fiji, gradually pushing northwards towards Hong Kong, which would hold out as long as possible, supplied by convoys fought through from Singapore, an idea which in the 1920s had been rejected as impossible to achieve with any measure of success. Once Hong Kong was relieved, Plan Y, essentially the second and third phases of the original War Memorandum (Eastern), would be implemented, with the battle fleet moving northwards, driving Japanese merchant ships from the seas, and provoking the Imperial Japanese Navy into the decisive naval battle. Dreyer was therefore very keen to see the defences of

Singapore completed, so that the storage of materials for Hong Kong and the other advanced bases could be commenced, as it seemed clear that the two bases were indispensable for naval operations against Japan.

The flaw with Dreyer's plan was that it still ultimately relied on a battle fleet being available for operations in the Far East and, as Backhouse was concluding, this was no longer a guaranteed certainty. Instead, Backhouse determined to refashion the Royal Navy's war plans to deal with German, Italian and Japanese threats, believing that it would be dangerously impracticable to send the bulk of the Royal Navy to the Far East for an unspecified time in the event of an Anglo-Japanese war. The Mediterranean Sea became the new focus of naval operations and, once the Italian military machine was broken down, that would be the time to gather sufficient forces to deal with either the Japanese or the German threats.[17]

As First Sea Lord, Admiral Backhouse brought Admiral Drax out of retirement to plan the future strategy and tactics for the Royal Navy. Backhouse had originally envisaged a fleet of four or five battleships going east, but Drax, who saw this as too small a fleet to fight with, envisaged a 'flying squadron' of two or three fast vessels being sent east to Singapore, forming the nucleus of a larger force to be sent out later. His strategy was to give the Mediterranean priority over the Far East:

> Drax favoured relying on small, fast, mobile forces to deter the Japanese from interfering and overrunning British interests in the Far East. Drax felt that a 'flying squadron' of two battleships or battlecruisers, an aircraft carrier, a cruiser squadron and a destroyer flotilla would be the ideal striking unit for this purpose.[18]

This force could be supplemented by an additional aircraft carrier to extend coverage to the Indian Ocean and Australasian waters and was intended for short-term operations only. Drax believed that this force would do a better job of defending British interests than a larger, slower force, securing British trade in the Indian Ocean and Malaya, hunting down any raiders as well as raiding Japanese lines of communication. He also suggested an additional eight cruisers, 17 destroyers, 15 submarines, two minelayers and 12 MTBs for trade and local defence. In Drax's revised war plans, it was envisaged that light naval forces, primarily cruisers and armed merchant cruisers, should be distributed along trade routes and other vital points to defend these from enemy attack. This would involve large forces, wide dispersal and the relegation of these forces to the defensive. Other forces, principally submarines and minelayers, would be located so as to contain an enemy fleet in harbour and follow them if they broke out, supported by the forces detailed for offensive operations against the enemy, the battle fleet.

The natural tendency would be to try and defend all possible trade routes with cruisers and possibly hunting groups formed around aircraft carriers and battle cruisers but, if this was done, then the forces available to form the battle fleet would be diminished. Drax believed that it made more strategic sense to concentrate force to deliver a knockout blow at a vital spot where the enemy was weak, and that with the likelihood of a war against Germany, Italy and Japan, Italy was the weakest of the three powers, and should be dealt with first.

It was therefore essential that the forces in the Far East be reduced to an absolute minimum until Italy was defeated or neutralized. The vital areas in the east were Singapore, Australia and the Indian Ocean, but Drax's proposed flying squadron was not powerful enough to deliver a 'knockout blow' against Japanese naval forces until either the arrival of a strong fleet from the Mediterranean, or, more significantly, US naval assistance in the shape of the US Pacific Fleet.

If the Japanese did attack Singapore, it had two options: a step-by-step attack or a direct attack. Of the two, the former had the advantage of consolidating during the line of advance and protecting Japanese sea lines of communication. A direct attack, however, would be swift and would catch the British off balance. The British aim was therefore to make the defences of Singapore as strong as possible to resist a direct attack and, in doing so, make it impossible to invest Singapore from the seaward side, whilst attacks overland, although not impossible, would be difficult in the absence of a secure base to work from.

Japan was expected to launch a step-by-step attack down the China coast, capturing Hong Kong first and moving to somewhere like Brunei or Sarao Bay on the west Indo-Chinese coast. This was still 500 miles away and Japanese lines of communication would be open to air, cruiser and submarine attack. So, provided Singapore had an adequate garrison, air forces and light naval forces, there was no need for a fleet to be permanently based there, and any naval forces sent to Singapore – Drax's 'flying squadron' – need be capable of no more than raiding Japanese lines of communication.

In short, the main requirements were to be fully developed defences and local naval forces capable of interfering with Japanese lines of communication. To this would be added a 'flying squadron' based on Singapore, adequate for the defence of Australia, and composed of battle cruisers or modernized *Queen Elizabeth*-class

DEVELOPING THE FAR EASTERN STRATEGY

battleships, two aircraft carriers, four large cruisers and eight destroyers. The ever-present threat of a strong reinforcing fleet joining this squadron to operate against the Japanese coasts, lines of communication and, eventually, their battle fleet would act as the necessary deterrent, whereas sending a slower, weaker but numerically larger fleet to Singapore would not be a deterrent to the Japanese. It would either be drawn out and defeated, or tied to Singapore as a 'fleet-in-being', thus being ineffective and lowering morale.

Backhouse had intimated this when he told Admiral Sir Ragnor Colvin, the Naval Member in Australia, in a letter in January 1939, that

> I regard any large scale expedition against Australia by the Japanese as most improbable so long as we have a fleet at all. It would be a tremendous undertaking on account of the distances ... attack on trade is a more likely operation, but there is a limit to what one or two ships can do in this direction. I think a more likely form of attack on the trade would be by cruisers and raiders and these your cruiser force should be able to contend with.[19]

Later on he added:

> Perhaps I might mention again the question of our sending capital ships to the East. At the present time we have none to spare, nor shall we have any in 1939. I do not believe there is the least chance of our having a row with Japan unless we first get involved in Europe, when possibly the Japanese might think the opportunity was good. Even so, we should be able to send some heavy ships to the East, not as many as we should like, but some, which should be able to safeguard our communications in the Indian Ocean and their presence would be sure to have an effect on possible operations by the Japanese against Australia or New Zealand.[20]

Backhouse also assured the Australian High Commissioner that, if necessary, seven battleships would be sent out to Singapore in the event of a war with Japan, although he remained doubtful about the likelihood of such an attack, or, if it happened, in Britain's ability to fulfil this obligation. In a letter to Vice Admiral Sir Percy Noble, the current Commander-in-Chief, China Station, written the same day as his letter to Colvin, Backhouse wrote that he saw no immediate need for a fleet to be sent to Singapore as a deterrent and that the Japanese would not attack Singapore, with their army so heavily involved in China.

Significantly, there were no British ships available, in any case, to form the British Far Eastern fleet. Four out of Britain's 15 capital ships were laid up for repair or refit, two others were due to go into refit and the others were needed in European waters to maintain a semblance of naval superiority. So, despite his reassurances to Australia, the reality was actually very different.

In Drax's reassessment of the Royal Navy's strategic priorities, the Mediterranean Sea, and actions against the Italians, were given the highest priority. Admiral Sir Dudley Pound, Commander-in-Chief, Mediterranean, certainly believed that the Royal Navy was the only one of the three armed services which could act offensively in the Mediterranean, even though he feared having ships and some bases subjected to almost constant bombing. For their part, the Italians felt that they had the capability to attack Malta and not Alexandria, which was beyond the range of their bombers in 1935.[21] In an assessment drawn up in November 1938, called 'The Strategical Aspect of the Situation in the Mediterranean on the 1st October 1938', Pound had advocated an early attack on Italian-held Libya from Egypt and French Tunisia. Such an attack would, he felt, have an effect on neutral countries such as Greece, Turkey and the Balkan countries in the Mediterranean, illustrating British and French resolve and possibly turning them into friendly neutrals; it would safeguard the Malta–Egypt sea route, releasing land, air and sea forces for operations elsewhere and obtaining superiority in the central Mediterranean Sea and Malta. Conversely, Italy would find it difficult if not impossible to reinforce the troops in East Africa.

However, Italy would not fail to make use of its geographical position and superiority in aircraft, and there was to be no question of allowing unescorted merchant ships or weak naval forces to operate in the Mediterranean. Any naval operation, however small, would require major defensive measures, as no ship would be immune to Italian air attack.[22]

This made the Mediterranean a unique theatre of operations. Once hostilities had begun, and until Italy was subjugated, the longer, but safer, Cape route would have to be used for supplying the eastern Mediterranean and Far East. And although the Mediterranean could be abandoned to provide the ships for Singapore, Pound recommended instead an early offensive in order to eliminate Italy from the war as soon as possible. But it was conceded in the Admiralty that the land and air forces were currently inadequate for such a campaign and that realistically the most that could be done was to hold up the expected Italian attack on Egypt and Tunisia. The Mediterranean Fleet would be kept in the Eastern Mediterranean, harassing Italian lines of communication and maintaining control of Egypt and the

Suez Canal.[23] If the worst came to the worst and Japan declared war, Pound was confident that the French Navy could tie down the remnants of the Italian fleet whilst the major British units were redeployed eastwards.

The whole strategical emphasis had now changed, and instead of a war in the Far East, war in Europe was expected to break out first, the Japanese declaring war merely to take advantage of Britain's preoccupation with this European war. Instead of being the fleet which moved eastwards, the Mediterranean Fleet was now to operate in the Mediterranean Sea, against Italian naval forces which could threaten lines of communication. In the place of large naval forces for Singapore, a much smaller squadron was envisaged.

Drax's emphasis was not so much on the size of the fleet but on its strength and mobility. But from this point comes the start of the debate regarding the composition of the force to be sent to Singapore in the event of hostilities, a debate which culminated in the sinking of the *Prince of Wales* and *Repulse* in 1941. Chatfield, a former First Sea Lord and now Minister for Coordination of Defence, was alarmed at the idea of a small squadron, fearing the political message this would give not only to Australia and New Zealand, but also to Japan. He pointed to the Anglo-French superiority in capital ships as a justification for being able to send a larger fleet to Singapore. Four or five British battleships, along with the seven French battleships would be more than adequate to deal with the Italian and German navies, whilst the seven to nine British battleships at Singapore, with their superior fighting qualities, would be more than a match for the Japanese.[24] Whilst Chatfield failed to convince the Strategic Appreciation Committee, he did reopen the issue in April 1939 when he challenged the Chiefs of Staff's view that only two capital ships could be sent to the China Station. A revised Chiefs of Staff assessment did concede that the seven battleships could be sent eastwards, but only if the Mediterranean was abandoned.

But it was expected that the threat of sending the British Main Fleet would be prominent in Japanese minds, and that this would act as a deterrent. Sending out a slower, weaker fleet would be no deterrent as it would either be drawn out and defeated by the Japanese or tied to Singapore, where it would be ineffective. What threat would the *R*-class battleships present to the Japanese, since they would be operating 3,000 miles from home waters and were recognized to be ineffective fighting units? The naval force should constitute a viable threat and this was more likely to come from fast battleships, three aircraft carriers (*Indomitable, Hermes* and *Ark Royal*) and the cruisers and destroyers that were proposed. Only such a fleet would be adequate to defend Australian waters and the Indian

Ocean. The arrival of the remainder of the battle fleet (and/or the intervention of the US Pacific Fleet) as reinforcements would seriously inhibit Japanese offensive actions. But if modern naval forces were not available, or if the Italians were not quickly defeated, or the Americans could not be persuaded to station battleships at Singapore, then the 'flying squadron' would become a hostage to fortune, offering a tempting target to the Japanese.[25] A deterrent only works, after all, if the opposition is deterred.

If this was the case, what was the size of any fleet that was to be sent to Singapore? Chatfield believed that losses in the Mediterranean should be accepted or the Mediterranean abandoned altogether rather than let the Japanese loose in the Indian Ocean. Thus a fleet of nine ships should be sent, leaving four in home waters to join with the French battleships in watching the German and Italian navies. Chatfield believed that not to send a fleet would risk the break-up of the Empire, and that at least it was better to lose the imperial possessions in fighting and not by default. Backhouse was not as sure and with the number of Italian ships nearly equalling that of the French, with the Japanese battleships all having been modernized, and with six ships needed in home waters to match the German expansion, only five ships could be spared to go eastwards, and not the battle fleet as promised to Australia. Vice Admiral Cunningham, Backhouse's DCNS and *de facto* First Sea Lord between March and May 1939 when Backhouse was ill, prepared a strategical appreciation in which he emphasized that there were so many variables that it was impossible to determine with any certainty the size of the fleet that could be sent eastwards. These included the strategic situation in home waters and the Mediterranean, Japanese strategy, US reactions and the role of the Soviets in the Far East. There was no question that a force would be sent, but how big a force and whether it could be done without affecting the situation in the Mediterranean were different matters.

Captain V.H. Danckwerts, the Director of Plans, added to the vagueness by minuting that if the United States were neutral at least four ships would have to be sent, but if they were an ally the Royal Navy could rely on the US Pacific Fleet, and therefore only two ships needed to be sent. If Italy were neutral, or neutralized, then strong forces were not necessary in the Mediterranean, and so seven ships could be sent out:[26]

> While it is not open to question that a capital ship force would have to be sent to the Far East, its composition cannot be forecast at present. In the most favourable situation at home and in the Mediterranean a battle fleet of 7 or 8 ships might possibly be sent. In other

circumstances it might only be possible to despatch a small force in the form of a 'flying squadron'. In the latter event it would be most desirable that capital ships with high speed and large endurance should be sent.[27]

Backhouse retired in 1939 and the new First Sea Lord, Dudley Pound, was unhappy with the concept of a 'flying squadron', believing that whilst it could possibly secure British sea lines of communication in the Indian Ocean and deter the Japanese from major operations in the South China Sea and around Australasia, if the Japanese did start large-scale incursions into these waters, then the British squadron would have to abandon Singapore and retire on Trincomalee. Consequently, he told Phillips, his new DCNS, to have all references to such a squadron removed from the war plan. He also revised his earlier opinion regarding the effectiveness of the 'knock-out blow' against Italy, as it could only work if both the army and the RAF devoted sufficient forces to it. Instead, he believed that the Royal Navy's Mediterranean Fleet could 'throttle' Italy out of the war, a process that could take a long time. In 'Strategy in the Mediterranean' written in May 1939, Pound suggested that although the only force capable of taking the offensive to the enemy in the event of a war with Italy was the Royal Navy, it could do no more than attack Italian sea lines of communication and naval forces, bombard ports and harbours, and use the Fleet Air Arm to attack and neutralize Italian warships at anchor. The Italians were expected to retaliate by bombing Malta and the Dodecanese Islands and any shipping that attempted to sail through the Mediterranean Sea, without encountering serious opposition. There was not much to retaliate with as it was unlikely that there would be any spare aircraft from the air defence of Great Britain, or from the bombing offensive.[28] Losses, of both merchant ships and warships, were inevitable.

When the Japanese blockaded Tientsin in June 1939, the strategic realities of the situation were brought home to the Admiralty as they were helpless to stop the Japanese from exploiting Britain's weakness in the region. There was too much to risk in Europe even to consider naval reinforcements eastwards, despite the risk of war. Economic retaliation was the only possible course of action unless the United States could be persuaded to join in any war. The Chiefs of Staff prepared a paper 'The Situation in the Far East' and were asked to prepare a second paper on the likely strength of a fleet to be sent eastwards, and especially on whether sending only two ships would be seen as a sign of weakness. Neither paper made happy reading. Despite judging the Imperial Japanese Navy as being some 80 per cent of the Royal Navy's efficiency, sending a fleet to the east could only be done by abandoning the Mediterranean, at a time when the situation in Europe was extremely fragile, and such a move inevitably would be interpreted as weakness by the neutral states of the Balkans.

The strength of the fleet to be sent eastwards had to be enough to secure Singapore, Australia, New Zealand and the sea lines of communication with India from Japanese attack and to support, by force, economic warfare against Japanese trade. But, also, a sufficient number of ships had to be left in home waters to guard against the German battleships. The Chiefs of Staff settled on a fleet of battleships for the Far East, leaving six out of the seven ships in home waters and sending the remaining ship to join the four from the Mediterranean, constituting a fleet of five ships to go eastwards. Only when *Revenge* and *Renown* recommissioned in the autumn could seven ships be available for Singapore. Japan was not expected actively to seek out a fleet action without provocation and so this fleet would have to conduct war against Japanese trade, hoping to draw major Japanese units south, possibly in cooperation with the United States Navy, and hoping to divide the main Imperial Japanese Navy battle fleet, which was expected to be standing on the defensive in case of US raids on Japanese possessions and even the mainland. Such assumptions took no account of US plans, such as War Plan Orange, however, and no serious joint planning seems to have been considered at this time.

As the situation in Europe deteriorated between 1938 and 1939 and Japan became increasingly aggressive, the Admiralty realized that their strategic plans and calculations of 1937–39 were no longer viable without the Americans, either by having the US Pacific Fleet stationed in Hawaii, where its presence and potential for action would inhibit Japanese aggression in the south, or as an ally. After the sinking of the USS *Panay* by Japanese aircraft, President Roosevelt and the Chief of Naval Operations (CNO), Admiral Leahy, had advocated closer cooperation with Britain, and Captain Ingersoll, USN, arrived in London in 1937 for talks with Captain Phillips, the DCNS. These talks were largely unproductive, envisaging only a strong US Fleet being based at Pearl Harbor and a British fleet at Singapore, but still with no thought of joint action. The Americans were wary of Britain's motives and preferred to support them in the Atlantic, freeing British vessels for the Far Eastern Fleet.

In 1939 the Joint Planning Committee felt that the Mediterranean was the theatre in which Britain was best placed to win or lose a war against Germany and Italy by striking at Italy first, and that a defeated or battered Italy would be a millstone around Germany's neck, although later Joint Planning Committee meetings felt that the situation was better with Italy as a neutral.[29] There were still some hopes that the US government would take a more positive role in checking Japanese expansion in China, allowing the British to concentrate on Italy in the event of war. The predicament facing the Chiefs of Staff in June 1939 was that only 11 capital ships were ready for sea, yet in order to meet the Japanese in battle the Admiralty reckoned that they would need at least eight capital ships, and even this was one less than the Imperial Japanese Navy possessed. Additionally, this would take virtually all of the available capital ships out of home and Mediterranean waters. Even though naval supremacy was not about just counting the number of battleships each navy had, and even though the Admiralty had, in the 1939 memorandum, attempted to quantify its assumption of superiority by rating the Japanese on a par with the Italians, it was obvious that the Royal Navy could no longer operate on two or three fronts without allies.[30] Only *Nelson* and *Rodney* could be considered as truly up to date, and they were nearly 20 years old. Three *Queen Elizabeth*-class battleships and the battle cruiser *Renown* were being modernized, and although modern cruisers and destroyers were joining the fleet, the Royal Navy had lost its early lead in naval aviation and was inferior to the Japanese in both aircraft carriers and carrier-borne aircraft. The Fleet Air Arm possessed only around 150 aircraft and, unlike the Imperial Japanese Navy, had developed a tactical doctrine that kept the aircraft carriers closely tied to the battle fleet, relying on the combined anti-aircraft fire of the fleet for protection from enemy air raids. Unlike Japanese aircraft carriers, which carried approximately 60 aircraft, including high-performance fighters, the aircraft carriers of the *Illustrious* class then being built had armoured decks, but could only accommodate about 36 aircraft, mainly strike and reconnaissance aircraft, with a few fighters to escort them to their targets, not to defend the fleet.

The outbreak of war in Europe seemed effectively to end any hope that the Royal Navy had of sending a battle fleet to the Far East, for the naval forces on the China, East Indies and Australian Stations could not be reinforced without taking valuable vessels away from others elsewhere. In fact, it was these stations which lost many of their warships as the Admiralty reinforced home waters and the Mediterranean, and the politicians would have to decide whether or not, for example, the Mediterranean Sea was to be abandoned to the Italians and the fleet sent eastwards. By 1941, when the decision was made by Churchill to send *Prince of Wales* and *Repulse* to Singapore, Britain had been fighting hard in the Mediterranean. Greece and Crete had fallen, Malta was enduring almost constant air bombardment, convoys were being fought through and Britain's 8th Army had been pushed back to El Alamein. Abandoning the Mediterranean would have seemed like an admission of defeat in the region, and certainly to Stalin in Russia.

As First Sea Lord, Pound stressed the importance of having the United States Navy assist in the defence of the Far East. In August 1940, after the fall of France and the entry of Italy into the war, he suggested that as the British were likely to be able to send only a battle cruiser and an aircraft carrier to Ceylon, the US Pacific Fleet, or at least strong detachments of the US Asiatic Fleet, be stationed at Singapore, serving both nations' interests. But the US Assistant Chief of Naval Operations declined, pointing out the logistical difficulties of operating a US fleet from Singapore, 6,000 miles away from the nearest American logistical base. US ships were also needed in the Atlantic on Neutrality Patrols. Making such a deployment was also politically unacceptable; the American public expected US ships to be defending the United States, not Britain's colonial possessions.

In October 1940, further staff talks were held, where Roosevelt's invitation to the British to tell him what they wanted gave a further opportunity to restate the case for concentrating a US force of eight battleships, three aircraft carriers, cruisers, destroyers and submarines at Singapore, virtually the whole US Pacific Fleet. The assumption was that, in a war against the Axis, the Royal Navy, already heavily engaged fighting the U-Boats and German surface raiders, would take responsibility for the Atlantic, and the US Navy the Pacific. After conversations with the First Sea Lord, Captain Kirk, the US Naval Attaché, noted:

> Singapore must be held at all costs ... It was an adequate fleet base from which allied forces could prevent the Japanese from extending their Asiatic theatre of war into the Indian Ocean ... With the US Fleet, or a substantial portion of it, based upon Singapore ... the Japanese Fleet will be contained north of the whole chain of islands comprising the Dutch East Indies.[31]

But such a move would be politically unacceptable to the still neutral United States, leaving their west coast undefended, and these proposals were rejected by Admiral Stark, the CNO. Instead, the United States preferred to concentrate, with the British, on the European theatre, holding the Malay Barrier until the war in Europe was won, and then concentrating in the Pacific against Japan. Whilst Churchill felt that this was a sound plan, and that the Japanese would not dare to go to war against a superior US fleet, the Admiralty was not so sure, and had no use for a plan that wrote off the Western Pacific. In December 1940, Phillips, DCNS, Harewood, ACNS(F)

DEVELOPING THE FAR EASTERN STRATEGY

Daniels, Director of Plans, and Admiral Bailey, Admiral Bellairs and Admiral Danckwerts discussed the US viewpoint and the necessity of trying to persuade them to station at least two battleships at Singapore whilst the Royal Navy sent *Renown* and an aircraft carrier to operate from Trincomalee.

The Admiralty were now more convinced than ever that US assistance was vital if they were to operate in the Far East, and an assessment of the Japanese Fleet's strength, made by the Director of Naval Intelligence, had the ten Japanese battleships equal to the six fully modernized Royal Navy ships. The other British ships, principally the *R*-class battleships, were really only useful for the defence of trade, not having the speed, armour protection or anti-aircraft batteries for modern war at sea. Inevitably, this had the effect of putting the Royal Navy at a distinct disadvantage in the Far East.[32]

On the eve of the departure of the delegates to the UK/US Staff Conference, the First Sea Lord tried again to persuade the Americans that only three US battleships needed to be based on the west coast of the United States if nine ships were at Singapore, and in the February 1941 Staff Conference in Washington the British again stressed the importance of holding Singapore, and endeavoured to persuade the Americans to base naval forces there. For their part, the Americans remained unconvinced, not believing that the Japanese would attempt to penetrate into the Indian Ocean and disrupt the convoys to the Middle East, as it was against the traditional Japanese strategy of drawing the enemy towards Japan for a decisive battle. If US ships were to be based at Singapore, this would possibly arouse Japanese fears and prompt aggression. Instead, the Americans favoured raiding Japanese trade routes and possessions, and reinforcing the Atlantic Ocean and Mediterranean Sea, allowing British ships to move eastwards to defend their imperial interests. Their Pacific fleet would remain at Pearl Harbor and

> it was by now apparent to the British delegation that it was improbable that they would succeed in getting the Pacific Fleet moved to the westwards to Singapore and that the Pacific fleet based in Hawaii would not pose a sufficient threat to Japan.[33]

In effect, then, British Far Eastern strategy, as laid out in War Memorandum (Eastern), was effectively dead. The Royal Navy seemed unable to provide a battle fleet and the United States, which could, was unwilling to use its battle fleet to defend the British Empire. Nevertheless, the US delegates did accept that Singapore was vital for the defence of Malaya and the Dutch East Indies but would not fall into line with British aspirations. It was apparent that the

Japanese moves to the south must eventually include an attack on Singapore and the Dutch East Indies, but the differences of opinion between the two nations meant that whilst the British wanted the US Pacific Fleet to be concentrated at Singapore, 6,000 miles away from Pearl Harbor, the Americans believed that concentrating there was the best way to curb Japanese expansionism, and, if it came to war, the best starting point from which to implement their own war plan, War Plan Orange. Like Churchill, the United States did not believe that the Japanese would abandon their traditional policy and move away from their home waters. So, in a minute to Churchill sent in February 1941, Pound accepted that if the Americans were not to be persuaded to deploy their fleet to Singapore, then a Royal Navy fleet must somehow be sent.[34]

The danger of the Japanese moving south was a key discussion topic at the August Atlantic Conference, but Roosevelt would still not commit himself to any action. However, on 25 August 1941, after Churchill suggested that, in view of the destruction of the *Bismarck* and with the United States Navy increasingly active in convoy defence in the Atlantic, reinforcements could be sent eastwards, the Admiralty felt that they could consider sending *Nelson, Rodney, Ramillies, Royal Sovereign, Revenge, Resolution*, one battle cruiser, one aircraft carrier, ten cruisers and 24 destroyers to Singapore.[35] Quantity – providing a deterrent – and not necessarily quality was the aim, although it is unlikely that this fleet would have been either much of a deterrent or have lasted long in combat with the Japanese. (When Admiral Somerville was faced with just this prospect with a similar force against Admiral Nagumo's aircraft carrier force in the Indian Ocean in early 1942, he was less than optimistic regarding his chances of survival.)

However, as the situation in the region deteriorated during 1941 there seemed to be realistic hopes of persuading the United States to offer some practical naval assistance, especially in stationing the Pacific Fleet at Singapore, as the presence of a strong US fleet in the central Pacific would, it was hoped, not only reduce the likelihood of any Japanese aggression, but also draw the United States more closely into an alliance with Britain against the Japanese. In addition, joint US, British and Dutch conferences also began to consider the use of Manila as a forward base, alongside Singapore and instead of Hong Kong.

Churchill, however, believed in a small, quality fleet in the region, operating from Singapore. He believed that the US Pacific Fleet would be tying down the Japanese fleet, and consequently he did not fear Japanese attacks on Singapore as much as attacks on shipping in the Indian Ocean by fast, detached Japanese squadrons. He wanted fast modern ships operating in the Indian Ocean and Singapore, capable of catching and overwhelming Japanese raiding cruisers, not the slow, *R*-class battleships. Mistakenly using the situation facing the Home Fleet as an analogy, he wrote to Pound with what he thought was a persuasive argument:

> *Tirpitz* is doing to us exactly what a *KGV* in the Indian Ocean would do to the Japanese Navy. It exercises a vague, general fear and menaces all points at once. It appears and disappears, causing immediate reactions and perturbations on the other side.[36]

At the same time, the Australian Prime Minister, Robert Menzies, wrote to Churchill, pressing for the strengthening of the region's defences in the face of an improved naval situation in the Mediterranean and the Atlantic. Menzies refused to be fobbed off with vague promises, but the ships were just not available to send in the numbers he wanted. The only way would have been to have taken ships from the Mediterranean to reinforce the Far East, and Pound believed that the Mediterranean was the decisive theatre. He believed that if the British lost control of the Mediterranean theatre of operations, then the Japanese government would seize their chance, declare war and the Far East would fall. But if the British kept control of the Mediterranean, they could at least come back to the Far East later, in conjunction with the United States. Under current circumstances, Pound wanted his best ships in home waters and the Mediterranean and all he was prepared to risk sending to Singapore were the slow, under-armoured *R* class.[37]

By October 1941 it was obvious that the Far Eastern situation had deteriorated and that war was only a matter of time. Japanese merchant ships were returning to home ports and reservists were being mobilized. In this climate the Admiralty was again asked if it was possible for ships to be sent to the Far East, and it again replied that a fleet of old ships could be sent. Churchill 'invited' the Board to consider sending out a *King George V*-class battleship and an aircraft carrier. It was obvious he would force the issue if he was again opposed and, reluctantly, the Admiralty agreed to the political value of sending a modern ship on a highly publicized voyage to the Far East, but did point out that it would take several weeks to assemble the necessary ships. Eventually, a group were ordered to sail to join Vice-Admiral Layton's China Squadron (until he was superseded by Rear Admiral Phillips, flying his flag in *Prince of Wales*). Force G, as it was known, arrived at Singapore on 2 December 1941.

Royal Navy capital ships were finally based at Singapore, but the force was far short of the battle fleet envisaged in the 1924 memorandum. Instead, there were just the two capital ships, a modern battleship and one old battle cruiser, four cruisers (three old and one modern) and a handful of destroyers, with the difficult, if not impossible, task of tying down Japanese forces by raiding trade and preventing Japanese incursions into the Indian Ocean, should war actually break out.

Earlier assessments of the Imperial Japanese Navy as being inferior had also been revised, and from thinking their ships badly built and top heavy, and their naval air arm as inferior, operating slow, sluggish, under-performing aircraft, on a par with the Italians, the British had gradually built up a different picture of the Japanese:

> Japanese warships, unlike the British and American, were designed exclusively for fighting. The navy tended to put heavier armaments on smaller hulls than the western navies, at the expense of protection. Generally sacrificed were living accommodation, defensive armament and radius of action in order to achieve maximum speed and offensive power.[38]

Phillips's force would have a hard job, even as a raiding force. When, in 1940, Churchill requested from the Director of Naval Intelligence a detailed list of Japan's capital ship strength, the reply, drafted by the then Captain Tom Philips, stated that the four refitted and modernized *Kongo*-class ships had no exact equivalent in the Royal Navy, but were superior to *Repulse* and equal to *Renown*. The four modernized *Fuso*-class battleships were at least equal to the modernized *Queen Elizabeth*s and the two *Nagato*-class battleships were equal to the two *Nelsons*.[39] The Japanese, therefore, had ten battleships at least equal to the fully modernized Royal Navy battle fleet, whereas of the 14 Royal Navy battleships, only seven, *Hood, Warspite, Valiant, Renown, Nelson, Rodney* and *Queen Elizabeth*, once her refit was completed, were capable of opposing the Japanese. *Malaya, Barham, Repulse, Royal Sovereign, Revenge, Resolution* and *Ramillies* were unmodernized, slow and only really fit for the defence of trade. And since this assessment had been made, *Hood* and *Barham* had been lost and *Warspite, Valiant* and *Queen Elizabeth* damaged. Thus, even with a modern battleship on station, the Royal Navy was at a severe disadvantage in the Far East and would be particularly hard pressed. In an impossible situation, strategically, the Royal Navy could only take comfort in the apparent balance of regional naval power.

Other allied naval forces in the Pacific area had also been built up slowly, and at the beginning of December, British, Free French, US, Australian and Dutch naval forces numbered 11 capital ships, three aircraft carriers, 36 cruisers, 100 destroyers and 69 submarines, opposing a total Japanese force of ten capital ships, six aircraft carriers and ten seaplane tenders, 36 cruisers, 113 destroyers and 63 submarines.[40] Even though this naval balance of power had been achieved, the odds were very much in Japan's favour. They had developed the largest battleships, the fastest torpedoes, the most advanced fighter, catapult planes in submarines, 18-inch guns and midget submarines. The Royal Navy only had Force Z on station and although the United States Navy had a total of seven aircraft carriers, only three of them were in the Pacific Fleet, whilst, until the arrival of *Indomitable*, or her replacement, the sole Royal Navy aircraft carrier east of Suez was the old, small *Hermes*. Allied naval forces were also widely dispersed, had no common signals or codes protocols, had not trained together and by no stretch of the imagination could they be regarded as a battle fleet. They were also on the brink of being decimated by bold Japanese actions at Pearl Harbor, Manila and off Malaya

Overall then, the Japanese had a clear, qualitative superiority. When *Prince of Wales* and *Repulse* were eventually sent to Singapore, it was primarily to allay Dominion concerns and as a sign of imperial resolve. They were meant to be a deterrent, and it was intended to reinforce the fleet at the earliest possible moment. Events overtook them.

NOTES

1. R. Higham, *Armed Forces in Peacetime: Britain, 1918–1940: A Case Study* (G.T. Foulis, London, 1962), pp. 136–7.
2. B. Bond, *British Military Policy between the Two World Wars* (Clarendon Press, Oxford, 1980), pp. 89–91.
3. Ibid., pp. 193–4.
4. Ibid.
5. Imperial War Museum, London, *Rear Admiral Ross Papers*, 86/60/1.
6. Moretz, *The Royal Navy and the Capital Ship in the Interwar Period*, p. 169.
7. Ibid., pp. 161–2.
8. R. Mallet, *The Italian Navy and Fascist Expansion, 1935–1940* (Frank Cass, London, 1998), pp. 7–37.
9. Bond, *British Military Policy*, pp. 218, 267.
10. Tracy, *Collective Naval Defence*, pp. 467–8, 475–9, for discussion on Hong Kong's use.
11. Cowman, *Dominion or Decline*, pp. 21–2.
12. Marder, *Old Friends, New Enemies*, pp. 26–7.
13. PRO CAB16/205, 'CID Strategical Appreciation Sub-committee: The Despatch of a Fleet to the Far East: Note by First Sea Lord', SAC4, 28 February 1939, p. 144.
14. Ibid., 'Memo by DCNS, 5 April 1939', pp. 237–41.
15. Ibid., 'CID 6th Meeting, 17 April 1939', p. 114.

16. Bell, *The Royal Navy*, pp. 80–3.
17. M. Murfett (ed.), *The First Sea Lords: From Fisher to Mountbatten* (Praeger, Westport, CT, 1995), p. 178.
18. Ibid., pp. 179.
19. PRO, ADM 205/3, 'First Sea Lord's Records, 1939–45: Letter to Sir Ragnar Colvin – Naval Member in Australia – from Admiral Sir Roger Backhouse, 9 January 1939', pp. 7–8. The Australian 8-inch cruisers *Australia* and *Canberra*, the three 6-inch *Perth* class, the older *Adelaide* and armed merchant cruisers would thus be adequate for defending Australia's mercantile trade.
20. Ibid., pp. 13–14.
21. Churchill Archives Centre, Papers of Admiral of the Fleet Alfred Dudley Pickam Roger Pound, 1877–1943, DUPO 4/4, ADM 116/3900, 'The Strategical Aspect of the Situation in the Mediterranean on 1 October 1938', p. 2.
22. Ibid., p. 3.
23. Bell, *The Royal Navy*, pp. 88–90.
24. Ibid., pp. 90–1.
25. Churchill Archives Centre, DUPO 4/4, ADM 116/3900, 'Strategy in the Mediterranean, 10 May 1939: Admiralty Reply to Pound, 6 July 1939'.
26. Marder, *Old Friends, New Enemies*, p. 54; PRO ADM 116/3863, Captain V.H. Danckwert's minute, 5 May 1939, approved by DCNS, 16 May 1939.
27. Marder, *Old Friends, New Enemies*, p. 58.
28. DUPO 4/4.
29. Bell, *The Royal Navy*, p. 65.
30. Marder, *Old Friends, New Enemies*, pp. 143–4.
31. Ibid., pp. 145–6, 'Notes on a Conversation between Captain Kirk, US Naval Attaché, and the First Sea Lord, 19 November 1940'.
32. Churchill Archives Centre, DUPO 5/2, 'Remarks by DCNS when Asked by Winston Churchill for a Detailed List of Japanese Capital Ships'.
33. Marder, *Old Friends, New Enemies*, p. 192.
34. Ibid., p. 191; M. Brice, *The Royal Navy and the Sino-Japanese Incident, 1937–41* (Ian Allan, Shepperton, 1973), pp. 90–1.
35. M. Middlebrook and P. Mahoney, *Battleship: The Loss of the* Prince of Wales *and* Repulse (Penguin, London, 1979), p. 27.
36. Ibid., p. 31.
37. Marder, *Old Friends, New Enemies*, p. 219.
38. Ibid., p. 230.
39. Churchill Archives Centre, DUPO 5/2, 'Remarks by Tom Phillips'.
40. Brice, *The Royal Navy*, p. 151.

5

Battle Fleet Tactics and a War in the Far East

> No captain can do very wrong if he places his ship alongside that of an enemy.
>
> Admiral Lord Horatio Nelson to his captains before the Battle of Trafalgar, 1805

The strategy, War Memorandum (Eastern), was also reflected in tactical developments in the 1920s and 1930s in so far as these were focused on a battle fleet action and the problems associated with this, especially handling a smaller battle fleet. This strategy was shaped to a large extent by the battleship and by the Washington Naval Treaty, leading such a fleet and obtaining a decisive victory.

Initially, British strategic and tactical assumptions still remained centred on a belief in the overall quantitive and qualitative superiority of the Royal Navy. Traditionally, the collision of fleets in battle identified the winner and the loser in a naval war, and thus who held command at sea. The 1925 Tactical Manual, for example, stated that the main principles of naval war were the maintenance of the objective, offensive actions, surprise, concentration of force, economy of force, security, mobility and cooperation with other arms, and that these, along with the organization, administration and supply of a battle fleet had to be borne in mind if the primary naval aim of a war plan, the destruction or neutralization of the enemy naval force, was to be achieved.[1] Formulating a strategy helped towards achieving this, but ultimate success was to be based on tactical superiority in battle. The strategy was to concentrate the battle fleet to enable it to do this, whilst the tactical function of all vessels was to impose maximum damage on the enemy battle fleet. Planning in War Memorandum (Eastern) began to reflect this, suggesting that the best, and probably the only, way of forcing a battle fleet action was by attacking Japan's sea lines of communication with cruisers, compelling its battle fleet to engage with the Royal Navy in a decisive battle:

> Throughout most of the inter-war period the objective of the Navy's blockade was to dislocate the Japanese war economy in order to weaken it in general terms, and ultimately, to induce the Japanese fleet to accept battle with superior British forces. Once the enemy fleet had been eliminated the Royal Navy would be in a position to enforce tighter control over Japanese trade and inflict decisive, economic pressure.[2]

As a consequence, the Royal Navy's tactics during the 1920s were developed with the aim of defeating the Imperial Japanese Navy, and avoiding an indecisive battle. War Memorandum (Eastern) initially concentrated on a decisive battle; the destruction of the Japanese battle fleet became the primary tactical aim for the Admiralty's tactical planners.

The most recent and, indeed, the only major battle fleet action, the Battle of Jutland, had highlighted several problems which needed to be addressed by the navy; principally, centralized control, gunnery and communications. Admiral Jellicoe, the British commander at Jutland, was a bureaucratic commander whose Grand Fleet Battle Orders covered some 200 pages and every likely contingency the Grand Fleet might encounter, whilst stifling initiative and subordinating commanders' own judgements.

Poor gunnery was another factor behind the disappointing results of Jutland, with rangefinders inadequate for battle ranges of 13,000 yards up to 18,000 yards, and, as a consequence, hit rates at Jutland had been as low as 1 per cent to 3 per cent. There was an obvious need for more effective fire control and a more effective rangefinder than the 9-foot Barr and Stroud.[3] There was also the need to coordinate tactics and gunnery, and to improve the liaison between gunnery officers and flag officers.[4]

Problems within communications had been highlighted at Jutland. Jellicoe's command style was signals dependent, Beatty's was not. Jellicoe had expected too much from the Signals Branch, who had to cope with lost aerials, vibration, gun and funnel smoke, and poor visibility. Training compounded the problems, with signallers trained at HMS *Victory* and wireless operators at HMS *Vernon*.[5]

When Beatty took over from Jellicoe the need for decentralization was recognized and a new signal – *MP* – was introduced for the Commander-in-Chief to use when he wanted to indicate that he was finding it difficult to control the battle fleet, and to remind squadron flag officers to act independently, whilst acting in close support of the

13. Retired Admiral, Percy Scott's 1920s prophetic view of a naval battle of the future, published at the time of the Bonar Law Enquiry. It was not so much that the Admiralty thought this view fanciful, more that they doubted the ability of aircraft to sink battleships unaided.

squadron to which the fleet flagship was attached.[6] In January 1918 the Grand Fleet Battle Orders were replaced by the two-page Grand Fleet Battle Instructions, the guiding principles to be observed by the fleet in contact with the enemy, and the Grand Fleet Manoeuvring Orders, detailing the handling of squadrons and flotillas. Significantly, these were instructions and guides, not orders. Beatty expected his subordinate flag officers to respond to the ebb and flow of battle.[7] The Grand Fleet was to remain together until the defeat of the enemy was certain, and attacks by individual squadrons were discouraged before then. Crippled enemy ships were to be left to the destroyers or submarines to deal with, and were not to be engaged by the battle squadrons. Further consideration of how best to handle a battle fleet in action continued for the remainder of the war and in subsequent years.

Whilst Jutland was not the decisive battle that many naval officers had hoped for and expected, the fact that it was not made it the centrepiece of study between the wars. Strategically, Jellicoe's handling of the Grand Fleet was successful and British naval supremacy had apparently been maintained without a decisive battle but simply by having a superior fleet. Tactically, however, there was much to learn, and within the context of War Memorandum (Eastern), with a smaller fleet, numbering perhaps no more than 15 battleships, tactics and command styles would have to be different, and the importance of initiative at all levels would become increasingly important.

As long as naval battle remained central to naval thinking and battle fleets continued to be built, a decisive victory remained the ultimate aim. Gunnery and fire control were, therefore, obviously central issues to be evaluated. Policy during the war had been to fire at maximum range and at a high rate of fire, overwhelming the enemy with a hail of shells. Post-war gunnery practices were aimed at improving basic proficiency, fine-tuning new techniques of firing and spotting salvoes, and concentrating the fire of several ships on one target, and various methods were tried to improve the performance of the Royal Navy's gunnery in battle.

Since August 1917, 'sequence' firing had been used, when pairs of ships fired, in turn, at regular intervals, each having equal opportunity to fire quickly and spot the fall of shot accurately, communicating changes of range and bearing either by wireless, flags or signal lamps. This system could also be used by larger numbers of ships, although the more ships employed, the slower the rate of fire would be, and the greater the difficulty of communicating.[8] In an attempt to increase the rate of fire, range clocks had been fitted to all capital ships and cruisers to improve the transmission of gunnery information, and the Vice Admiral commanding the 5th Battle Squadron in 1917 reported:

> the visual signalling clock dial is considered superior to lamps for intercommunication of ranges because 1) the indications on these dials are always showing, whereas signals made by lamps must be received at a particular moment or they are lost altogether and 2) they eliminate mistakes.[9]

Transmitting this information from ship to ship ensured a concentration of fire on targets and in 1918 the Grand Fleet was a much more capable force in terms of gunnery than it had been at Jutland. The 1919 Admiralty War Game Concentration Factors calculated an improvement of 2.25 per cent for a pair of battleships or battle cruisers, and 4.5 per cent for a division of four ships, firing this way, over a single ship firing.[10]

Accurate gunnery still remained fraught with difficulty, though, and fighting at close ranges still seemed to offer the best chance of achieving destructive hits and not wasting ammunition, even though it was not always as efficient as longer range. Opening fire at a longer range, with the aim of disrupting an enemy's deployment and maybe scoring a lucky hit, had the advantage of giving the shell a higher trajectory and a steeper flight, with a greater chance of penetrating deck armour and causing a critical hit, but had the disadvantage of being dependent on spotting by aircraft, and the spotter would be vulnerable to attack by enemy fighters, or might not be able to get airborne in bad weather. Close-range action, by comparison, could be more easily observed, but was less effective against armour plate, although the ship's ability to fight might be seriously affected as the bridge, upperworks and gun turrets were destroyed.

Further improvements continued to be made to gunnery and fire control. Many ships were re-equipped between 1917 and 1923 with longer-based rangefinders, with better mountings less affected by vibration. So by 1921 the Royal Navy was able to start to shift to 'master ship' firing, with a single ship controlling the range and deflection and all ships in a squadron firing a salvo approximately every 20 seconds.

Improvements were also made to ships' fire control. The Dreyer Table, an early computer, intended to replace the earlier *dumaresq* system of fire control, was fitted to most of Britain's battleships and was intended to provide a more complete solution to working out all of the variables between course, speed and bearing of target, although several *dumaresq* were still in use, especially in the older battleships and armoured cruisers. Dreyer's refinement allowed the Transmitting

Station in the ship to take observations from the spotting top and originally a single rangefinder, although this was improved during the First World War to allow plotting of ranges from multiple rangefinders.[11] However, the manual operations of the Dreyer Table were slow and fire control personnel found it difficult to keep their instruments adjusted to the correct ranges and bearings, leading to a slow rate of fire, although the introduction of the Admiralty Fire Control Table (AFCT) promised to rectify the shortcomings in gunnery, and allowed the 'master ship' to coordinate the shooting of up to four others to improve concentrated firing and spotting on a single target.

Trials of a successor to the Dreyer Table, the Admiralty Fire Control Table, Mark I, began in 1925, and were a complete success. Once the firing ship had observed the present range and bearing of an enemy ship and estimated enemy course and speed, the computing mechanism (the AFCT) determined the aiming point (effectively the future position of the target when the shell landed), as well as its range and bearing from the firing ship at the moment of firing. Range was then transformed to the required elevation angle of the guns by the Range to Elevation Gear. The Table already had a number of known variables programmed into it, and a series of constants, including own course and speed, wind strength and direction and drift. Variables would include enemy course and speed, range and bearing. The number of previous firings made by each barrel would also be calculated in, as these would affect the shooting of each individual gun. Information on the enemy was passed from the Gunnery Director to the Transmitting Station, where the AFCT computed the exact range, bearing, speed, deflection and elevation back to the Director Control Tower and the guns. The operators in the turrets set the elevation and train, following the Training Models from the Transmitting Station, and when the circuits were made and all of the 'Gun Ready' lamps went on, the Gunnery Officer in the Director pulled the trigger and fired off the salvo; C.S. Forester provided a vivid description of the workings of a later mark AFCT in his novel, *The Ship*:

> The three rangefinders in the ship were at work in an instant ... Down in the Transmitting Station, a machine of more than human speed and reliability read off all three recordings and averaged them. Each of the other observers in the Director Control Tower was making his particular estimate and passing it down to the Transmitting Station, and down there, by the aid of these new readings, the calculation having been made of how distant [the target] was at the moment, other machines proceeded to calculate where [the target] would be in fifteen

seconds time. Still other machines had already made other calculations; one of them had been informed of the force and direction of the wind and would go on making allowance for this automatically ... each gun had been given its individual setting to adjust it to its fellows. Variations in temperature would minutely affect the behaviour of the propellant in the guns ... so one machine stood by to make the corresponding corrections; and the barometric pressure would affect both the propellant and subsequent flight of the shells – barometric pressure, like temperature, varied from hour to hour and the Transmitting Station had to allow for it. And the ship was rolling in a beam sea – the Transmitting Station dealt with that problem as well.[12]

Shell splashes would be observed, and the results passed to the Director Control Tower and the Transmitting Station. In accordance with the average, the elevation of the guns was adjusted up or down the range ladder, and the Gunnery Officer also used his own judgement to make alterations. Constant training was vital to keep this whole system working effectively and to keep firing off salvoes. About ten or 12 men in the Transmitting Station operated the Table itself, usually Royal Marines, as they were regarded as being more used to the strict discipline vital for working it efficiently.[13]

Models and derivatives of the AFCT Mark I were fitted to *Nelson* and *Rodney* as they completed in 1927, and to *Warspite, Queen Elizabeth, Valiant* and *Renown*, as they were refitted in the latter half of the 1920s. Older ships (*Barham, Malaya, Repulse, Hood* and the R class) had later model Dreyer Tables until their refit, and it was expected that in any battle the older ships would be in squadrons, with the refitted ships acting as the 'master ship', passing information to the squadron using their Range Clocks fitted to the fighting tops and the mainmast. It was anticipated that 'master ship' firing would take place at the start of an action, when communications were intact, and then shift to paired firing once the action had been joined and the range had closed sufficiently to lessen the discrepancies between the Dreyer Table and the AFCT.[14] But even by the outbreak of the Second World War, only the new battleships of the *King George V* class, along with the *Nelson, Rodney, Warspite, Valiant, Queen Elizabeth* and the battle cruiser *Renown* had the AFCT; the *Barham, Malaya, Royal Oak, Revenge, Royal Sovereign, Ramillies, Resolution* and the battle cruisers *Hood* and *Repulse* all still had the older Dreyer Tables.

Gunfire needed to be rapid, accurate and concentrated if it was to be a compensation for the smaller numbers of ships. Opening fire at a decisive range, at as close a range as was possible, as opposed to firing at long range, would overwhelm the enemy with a hail of shells and would, it was felt, go a long way to achieving a moral ascendancy. For example, when Rear Admiral Cowan, Commander-in-Chief on the North America, West Indies Station, wrote to Admiral Roger Keyes, in November 1926, he stressed this, believing a well-drilled ship or squadron would be able to close with an enemy before opening fire, without the need to bother with long-range, plunging fire, which both wasted ammunition and made spotting more difficult earlier on in the battle: 'You want both to consternate the enemy by the sight of their leader drifting past the whole line broken and disabled – and to inspire and encourage your own lot by the same panorama.'[15] Here discipline, morale and *esprit de corps* were equally important as effective gunnery, since closing to close range would inevitably result in damage. However, with the Royal Navy's tradition of aggressive, close action, this was a feature that was going to become commonplace in tactical doctrine.

As well as these developments, another sign that the Royal Navy wanted to profit from the lessons of the First World War was the setting up of establishments such as the Naval Staff College, at Greenwich in June 1919, and the War College, also at Greenwich, where Richmond made a significant impact on the curriculum with his study of tactics and tactical developments. There was also the Tactical School, founded at Portsmouth Dockyard in 1924, at the suggestion of Admiral Dreyer, to provide a more scientific study of tactics. Here, Madden's Atlantic Fleet Battle Instructions were developed into the Battle Instructions, a common set of instructions to be issued to the entire navy. In addition, tactical problems were discussed and worked through on the tactical board, including those associated with the Battle of Jutland. Tactical experimentation was also studied in sea exercises and, as a result of these exercises, the Naval War Manual was produced, which aimed at keeping the battle fleet as efficient and effective as possible.[16] From the surviving copies of the Naval War Manual, Progress in Tactics, the Battle Instructions and the writings of officers such as Drax and Richmond, it is possible to analyse some of the tactical developments, and to examine how a fleet would have been expected to fight if it was sent eastwards.

The poor performance of the Royal Navy during the First World War was partly due to a lack of initiative, and an overdependence on written instructions, and it was felt necessary to move away from detailed written orders in the future, as these could become an inhibiting factor to decisive victory.

Drax, who later was to play a prominent role, under Backhouse, in recasting the Far Eastern strategy, had been thinking about Battle Instructions since 1916, when he was a captain with the Grand Fleet, and he felt that they should be brief, not voluminous, manuals full of complex instructions. His Grand Fleet Battle Orders, of December 1916, were based on eight principles: cutting the enemy off from his base; not engaging until the whole force could open fire simultaneously; having a strong, fast van; crossing the enemy's 'T' concentrating fire on the enemy's van and then, as the enemy draws into range, concentrating fire against all targets in range; using superior speed to reach advantageous positions and allow the fullest freedom of movement; dividing one's squadrons into slow and fast groups and inflicting an early, simultaneous blow; firing from extreme range with destroyers and aircraft attacking early to disrupt the enemy's deployment.[17]

In January 1920 Drax gave a series of six lectures to the Naval Staff College, Greenwich, on the subject of tactics.[18] He pointed out that as it was unusual for two fleets to meet at a time and place when both wanted to fight; one fleet would usually be trying to retire. So, when two fleets did meet, a concentration on the enemy's van or rear would cause severe problems for this fleet, virtually forcing it to accept battle on unfavourable terms. To do this, though, would require fast ships and, therefore, dividing the British battle fleet into fast and slow divisions seemed advisable. In the battle fleet of the 1920s, the *Queen Elizabeth*-class battleships and the battle cruisers *Hood*, *Renown* and *Repulse* would have the combination of the speed and effective fire control equipment to be capable of acting in this way, although aircraft, which had the speed and the punch – the airborne torpedo – could also fulfil this role. When they joined the Fleet in the mid-1920s, the two *Nelson*-class battleships, 2 knots slower than the *Queen Elizabeths*, but fitted with the modern AFCT Mark I, would be able to act as 'master ships' for the *R*-class and *Iron Duke*-class battleships. In battle, once the enemy's battle line had been disordered, the fast division could attempt to manoeuvre either across the enemy's bow or stern to bring the action to a decision.

In a further attempt to abandon the rigidity of the wartime Battle Orders, Drax also pressed for every weapon to be used to defeat the enemy battle fleet, with all ships firing on all enemy ships, not just engaging their opposite number in the line of battle, and not just fighting their own types, a situation he likened to 'dog eat dog'. Instead, battleships were to engage any other vessel coming within range, whilst the battle cruisers were to support the cruisers in the scouting line, drive in the enemy scouting line, form a fast division, cooperate with battleships, and engage any weaker vessels coming within range. Light cruisers were expected to scout and screen their own fleet, attack the enemy line with torpedoes and repel similar attacks by enemy vessels. Destroyers had the responsibility of forming an anti-submarine screen and attacking enemy submarines, as well as attacking the enemy battle line with torpedoes, repelling similar attacks by the enemy destroyers, and laying gas clouds and smoke screens, whilst fleet submarines were expected to attack enemy aircraft carriers, lay gas clouds or smoke screens and sink disabled ships. Aircraft were also given a full part to play. Their roles included scouting, offensives against enemy aircraft, torpedo attacks against the enemy line or aircraft carriers, making machine gun attacks on exposed personnel, spotting and laying gas clouds or smoke screens.[19] The enemy fleet would, in short, be overwhelmed by the sheer ferocity of attack, fought at close range and impossible for an admiral to direct centrally.

Plans would therefore have to develop quickly and, in a return to 'the Nelson Touch' (or possibly, given the First Sea Lord's reputation, 'the Beatty Touch'), all flag officers would have to know the intentions of their admiral beforehand, as signalling would be kept to the barest minimum. Such a style exactly suited Beatty's personality and image, and with him as First Sea Lord, and so many of his supporters in positions of influence, the climate was right for change. But there was no doubt that with the fleet fighting in this way to annihilate the enemy, it was a high-risk scenario, one in which the personality of the admiral was going to be the outstanding factor in any future battle. Success or failure in battle was going to be the result of constant training and the admiral's efforts to instil aggressive action into his subordinates, and not just by a study of manuals and regulations.[20]

This return to a more aggressive approach to tactics had a great appeal for the Royal Navy, especially after the disappointing lack of a decisive battle fleet action in the Great War. Nelson and Beatty were both perceived as extremely successful and charismatic fighting admirals, whose approach to battle suited both the Royal Navy's self-image and the belief in decisive fleet battle. Increased technical sophistication and expertise did not lessen their appeal; the Royal Navy's 'collective memory and institutional ethos' confirmed the success of aggressive leadership and after the self-doubt promoted by the Great War, many officers embraced this new emphasis with enthusiasm.

Inevitably, without some overall guidance this increased reliance on initiative might have led to differences in tactical training, but by having one set of Battle Instructions and coordinating joint annual manoeuvres, the Mediterranean and Atlantic Fleets could each study and benefit from the tactics of the other, whilst smaller fleets and detached squadrons would be able to benefit from the official Admiralty publications which reviewed gunnery, tactics, the combined fleet exercises and war readiness. So, a system, of sorts, developed:

> Each CinC was to devise his own practices, analyse them and record them and their lessons for the immediate benefit of his Fleet. The ACNS would receive the firing reports, and, acting as a central control, would circulate the important practices and the lessons learnt to all fleets; so keeping the scattered fleets and squadrons in touch with each other, and giving the small forces, whose opportunities for training were often small, the benefit of the lessons learnt and the alterations in methods adopted by the two main fleets. The Admiralty could also order any fleet to carry out during the year any form of active they thought necessary.[21]

Constant evolutions and constant exercising would keep ships and crews highly trained, allow flag officers and captains to get used to using their own initiative, and keep abreast of contemporary developments in the Royal Navy, and in foreign navies. Success or failure in battle was going to be the result of constant training and offensive, aggressive action, not the result of studying manuals.[22] In the post-Washington world, with a smaller battle fleet, this meant a more aggressive doctrine with the aim of destroying the enemy's cohesion by manoeuvring and concentrating gunfire to overwhelm a portion of the enemy, using as many tactical initiatives and weapons as possible.

Richmond expanded on this theme in further Commandant's Lectures he gave to the Naval Staff College. He, like Drax, saw the fundamental tactical principles to be the object, the offensive, surprise, concentration, economy of force, security, mobility, and cooperation.[23] The object was clearly defined, as it was always the destruction of the enemy force, achieved either through decisive battle, or through the destruction of trade and the enemy's will to fight.

At the basis of Richmond's writings lay the belief that there was only one sound, guiding principle: 'Victory First'. He was, not unnaturally, supportive of the revisions in the 1922 Tactical Manual, and the Naval War Manual of 1925. Both of these documents reflected the main tenets of Drax's and Richmond's ideas about increased decentralization, the Tactical Manual, for example, stating:

> As a general principle the main fleet should keep together until the defeat of the enemy has been ensured. Attacks by a division or squadron, separately, on a portion of the enemy should, until then, be avoided; nor should ships be detached from the battle fleet to deal with

enemy disabled ships until the defeat of the enemy fleet is certain. Disabled ships should be attacked by submarines.
An exception to the plan may occur:–

   a) If the enemy plan of battle is one of tactical division

   b) If the rear is not in range and the Admiral considers it improbable that the general conditions will allow of his cooperating with the main body of the fleet. In this case he might detach a portion of his squadron to deal with enemy ships which fall out.

The detached force should not be stronger than is required for the duty. The remainder of the rear squadron should press on and use every endeavour to get into gun range of the main body of the enemy.[24]

Tactics was the means of achieving this end. But it was a means, not an end. Formations, orders and battle instructions were not to be considered as the only constituents of a study of tactics. Instead, an admiral should concentrate his forces and overwhelm an enemy, using mutually supporting divisions to isolate and attack one isolated enemy division after another by a superior concentration of gunfire. Once the enemy fleet was scattered and disabled, then the attacking fleet could divide into battle squadrons and complete the destruction. The centralized control as used by Jellicoe, at Jutland, would be ill suited to these fluid conditions and naval warfare was no longer dependent on the single line of battle. An emphasis on manoeuvre warfare at sea was necessary.[25]

Like Drax, Richmond was a traditionalist, not a materialist, and saw war as an art, and success in war dependent more on moral than physical qualities, on spirit rather than numbers of ships:

> Neither numbers, armament, resources, nor skill can compensate for lack of courage, energy, determination and the bold, offensive spirit which springs from a natural determination to conquer. The development of the necessary moral qualities is therefore the first object to be attained in the training of the Navy. Next in importance are organisation and discipline, the training of the mind and body, and the proper use of weapons. The final essential is skilful, resolute and understanding leadership. Unless the plan of action is sound and has been carefully prepared the highest moral and physical qualities on the part of the personnel will be unavailing, while, once battle is joined, success or failure will depend largely on the tactical skill and initiative of subordinate leaders.[26]

The art came in assessing each situation; quick judgement and decisive thinking had to be cultivated and encouraged at all levels.[27] The primary objective of any naval war plan remained the destruction or neutralization of enemy naval forces and all of the British battle fleet was to work together with one aim: the destruction or neutralization of the Japanese battle fleet. This could not be done until all subordinates fully understood their admiral's intentions, and control in battle was decentralized. Control by the admiral was limited to controlling the fleet during the approach to battle, disposing of the vessels in such a way as to gain the optimum positions for the deployment of guns, aircraft and torpedoes. The subordinate flag and squadron officers controlling their own divisions would relieve the admiral of the necessity of constantly signalling. Victory was only possible by offensive action and high morale; good leadership and organization, *esprit de corps*, patriotism and discipline were all contributing factors to this.

The Tactical Manual also stressed the need for concentrating the battle fleet whilst in action until it was clear that the enemy battle fleet had been defeated. Until that time, detaching battle squadrons to act independently would weaken the power of breaking down the main enemy resistance and achieving victory. Concentration of fire on individual enemy ships would quickly break up the cohesion of the enemy line, disable the enemy battleships and allow for greater concentration on the rest, gradually increasing the weight of fire on individual enemy battleships as more were disabled or sunk.[28]

As sea battles were more than just a straightforward engagement between ships of equal speed and firepower, tactical training, using initiative, exercising and a thorough knowledge of doctrine were to form the bases of fleet tactics in the 1920s, and such a philosophy was well suited to the post-Washington battle fleet, which could no longer rely on overwhelming numbers in battle, but had to develop into a highly trained force. So, all ships and weapons had to be used aggressively to achieve victory, and voluminous instructions and reliance on the 'all seeing, all knowing commander-in-chief' had no place in the smaller battle fleet.

The application of all of these ideas was seen perhaps in the 1922 Atlantic Fleet Battle Instructions, originally written when Sir Charles Madden flew his flag as Commander-in-Chief. Madden is not usually placed in 'the Beatty camp', but he was pragmatic enough to recognize that the changing situation after the war needed a different tactical approach. His Atlantic Fleet Battle Instructions became the principal guidelines for a fleet fighting a battle during the decade of the 1920s. Revised in 1925, and again in 1927, when they were adopted throughout the fleet, first as Battle Instructions and then, with some revisions, as Fighting Instructions, they gave a clear indication of just how the Royal Navy would have fought the Japanese.

For example, fully aware of the potential of the naval air arm after a 1919 exercise raid on his fleet at anchor in Portland Harbour, Madden saw a key role for aircraft of all types in his Battle Instructions. He proposed a total of 11 aircraft carriers: three for reconnaissance aircraft, four for spotters, two for torpedo bombers, one for anti-submarine aircraft and one for fleet defence fighters. His 14 capital ships and ten cruisers would also carry fighters or spotters, whilst he envisaged four more spotting carriers for the ships commissioned from Reserve and a further two for the Mediterranean Fleet. Most of these were to be small aircraft carriers, merchant ship conversions similar to *Pegasus*, with the exception of the reconnaissance carriers which would be converted from the large light cruisers *Furious, Glorious* and *Courageous*, able to carry a large air complement and operate wheeled aircraft. Madden reckoned that it would take at least three years to provide the aircraft and their carriers, but it would be one step in setting up an air policy for the fleet.

Aircraft were also expected to give the first information on the enemy's presence, later confirmed through direct contact by cruisers,

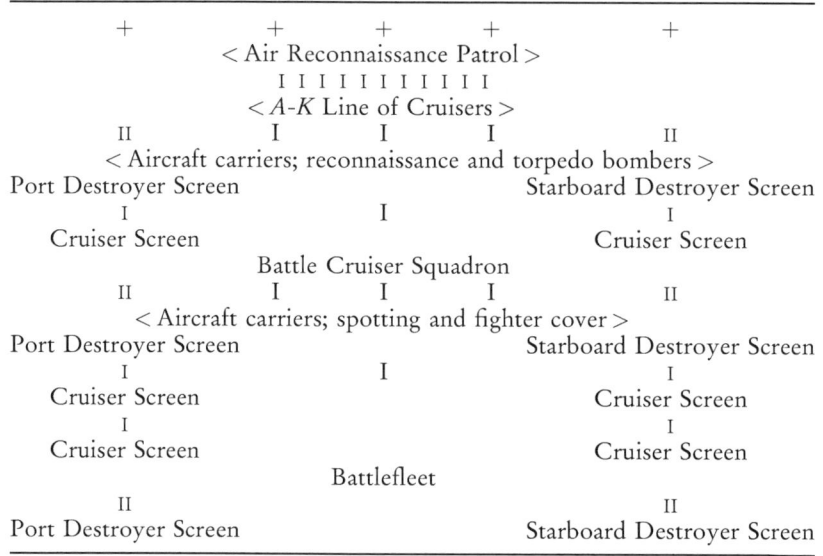

Table 3
British Battle Fleet in Cruising Formation, 1924

```
        +           +        +        +              +
                  < Air Reconnaissance Patrol >
                     I I I I I I I I I I
                      < A-K Line of Cruisers >
        II            I        I        I            II
              < Aircraft carriers; reconnaissance and torpedo bombers >
Port Destroyer Screen                      Starboard Destroyer Screen
        I                                              I
  Cruiser Screen                                 Cruiser Screen
                       Battle Cruiser Squadron
        II            I        I        I            II
              < Aircraft carriers; spotting and fighter cover >
Port Destroyer Screen                      Starboard Destroyer Screen
        I                    I                         I
  Cruiser Screen                                 Cruiser Screen
        I                                              I
  Cruiser Screen                                 Cruiser Screen
                             Battlefleet
        II                                            II
Port Destroyer Screen                      Starboard Destroyer Screen
```

Source: Till, *Air Power and the Royal Navy*, p. 138.

supported by battle cruisers.[29] Then, combined attacks by air and surface forces were expected to break up the enemy's cruiser screen and support torpedo attacks by the destroyers.[30]

As the battle fleet was vulnerable to aerial detection and attack, an air striking force would be launched immediately the enemy fleet was sighted, with their objectives, in order of importance, being enemy aircraft carriers and enemy aircraft, either in the air or in the act of launching:

> Fighting aeroplanes and torpedo aeroplanes organised as an air striking force are to attack enemy aircraft carriers and aircraft at the earliest possible moment. The attack will be launched by the Admiral's order. Subsequently if torpedo aeroplanes are available for other purposes they will be used against special objectives by the Admiral's order.[31]

Carriers would be attacked by torpedo bombers and by fighters dropping light bombs and machine gunning the flight decks.[32] Once the sinking of the aircraft carriers had deprived the enemy of their long-range strike and reconnaissance capability, the battle fleet could close and commence a surface action. Other aircraft would be spotting, shadowing the enemy fleet and keeping enemy submarines submerged, so that they could not use their surface speed to take up attacking positions. Aircraft would also be used, along with destroyers and cruisers, to disrupt the enemy battle fleet forming its own line of battle.[33] Once the cruisers and battle cruisers made contact with the enemy screen,

> it becomes the primary duty of the British battlecruiser squadron to attack and destroy them. If the enemy battlecruisers or fast divisions are supporting their light cruisers, they should be engaged closely and, if possible, prevented from rejoining their main force; the British light cruisers at the same time are to press on with the object of locating the enemy main force ... If, during the course of the battle, the enemy breaks off and retires, it is the duty of the battlecruisers and light cruisers to regain and maintain touch with the enemy, and the Flag Officer Commanding Battlecruiser Squadron, is to take command of the light forces at the van in order to carry out this important duty.[34]

With the enemy battle fleet's deployment already in some confusion due to the attacks by aircraft, battle cruisers, cruisers and destroyers, the British admiral would now attempt to place the whole of his battle fleet in the best position to engage, aiming to deploy at right angles to the line of bearing of the centre of the enemy fleet, battle squadrons being staggered to present a more difficult target whilst bringing the maximum broadside to bear. The staggered disposition would also allow the admiral's signals, especially the fire distribution plan, to be more easily read. Accurate and constant reporting of the enemy's course, speed and position by aircraft and cruisers was also going to be vital in this stage.

As soon as the enemy battle fleet was sighted, the British admiral would signal the fire distribution plan, which would identify which part of the enemy line was to be engaged. If this signal could not be made, then the squadron flag officers would fire at their opposites, leaving no ship that was in range unengaged. British battleships would open fire at maximum range, firing rapidly as the range closed. When visibility permitted, the action would be fought within 20,000 yards, the extreme range of 13.5-inch guns, but outside 15,000 yards, the accepted range of most torpedoes at 25 knots. The vanguard were not to approach nearer than 16,000 yards and, if the British battle fleet had not deployed by this time, the van squadron was to turn away and remain outside torpedo range.[35]

The whole fleet might not come into action simultaneously, in which case divisional commanders could open fire without specific orders from the admiral and, if individual ships sighted the enemy before their divisional flag officer, they, too, need not wait for orders to open fire.[36] Naval battles unfolded quickly and once battle was opened the admiral could not control the fire distribution of the fleet and the fire control guidance was that each division, aided by attendant aircraft, should aim to deliver an overwhelming concentration of fire on one of the enemy ships as soon as it came within maximum range, moving from long-range, 'master ship' firing, to 'paired' firing, aiming to overwhelm an enemy with accurate salvoes:

> The great range and large bearing arcs of modern guns render a gunnery concentration by tactical means very difficult to attain, and even if attained, the mobility of a fleet is so great that the enemy can quickly extricate his van or rear from a position of tactical disadvantage.
>
> Since concentration of fire is still a necessity to ensure the quick destruction of enemy ships at long range, or under difficult conditions of visibility, it is important that full advantage is made of any opportunity for effecting a gunnery concentration ...[37]

Concentrating on the head of the enemy line was deliberately done to make the enemy's deployment more difficult, and to lessen their morale, by hitting them whilst they were unable to hit back. As the range decreased and more enemy ships started to open fire, concentration was to be lessened to engage these enemy ships, ensuring all enemy ships were engaged but taking any suitable opportunity to increase the concentration on any one vessel, for example, when smoke obscured parts of the enemy line, reacting with the vigour that these Battle Instructions constantly advocated. The Fleet Flagship would control its own division in battle, not the whole fleet, whilst other divisions would be controlled by their respective flag officers:

> The line of battle, when formed, is to be considered as consisting of a number of squadrons and divisions, the flag officers commanding which have full authority to manoeuvre them so as best to employ their gun and torpedo armament, subject to the maintenance of close support and cohesion between them, without which the enemy cannot be defeated.[38]

Once the battle was under way, the admiral's control would necessarily be limited, and his subordinate flag officers were expected to be prepared to control the battle in their immediate vicinity, deciding on their own fire distribution and movements, rather than await orders which the admiral may not have been able to issue, either through ignorance of events, or because exercising control over all of his forces was impossible in the heat of battle. The battle squadrons would endeavour to stay in close support of one another until the defeat of the enemy was ensured, and only then were squadron commanders to consider separate attacks by their battle squadrons against a fleeing enemy:

> (a) Squadrons and divisions to remain in close support and maintain the cohesion of the whole battle fleet until the defeat of the enemy is assured.
> (b) To keep the enemy engaged, to allow him no interval to recover from a disadvantage and to overwhelm him as quickly as possible at decisive ranges.
> (c) The range to be generally inside the range of the shortest ranged gun of the main armament of the battle fleet.
> (d) The action generally to be fought outside enemy torpedo range, but the Admiral may decide to accept the torpedo menace.[39]

Points (c) and (d) are of great significance; the fleet should all be engaged, but, unless there was no alternative, the battle fleet should not be exposed to the risk of a torpedo attack. However, if these were made, they would be countered by turning towards the firing ships, maintaining contact, and not away, as Jellicoe had done at Jutland. At this stage of the battle, the fast division of battleships would be steaming ahead to try to turn the enemy back towards the main line of battle; battle cruisers, cruisers and destroyers would be attempting to attack the enemy screening vessels with gun and torpedoes; aircraft would be launching attacks; submarines torpedoing crippled enemies; and the major portion of the battle fleet would be steaming to complete the destruction of the enemy fleet. The admiral retained command of the whole fleet until victory was assured. In keeping with their role as fast scouts, however, the battle cruisers were allowed some further discretion:

> Failing instructions from the Admiral, the flag officer commanding the battlecruiser squadron must form his own judgment to which flank of the battle fleet he should proceed during the approach ... He should keep within supporting distance of the main fleet, taking care not to come under the fire of the enemy battle fleet before it is fully engaged by our own.[40]

If attacked by enemy aircraft, submarines and destroyers, the battle fleet must continue to engage the enemy, relying for defence, first, on screening vessels and aircraft, then on secondary batteries, and then, only when necessary, on turning away from, or into, torpedo tracks, keeping the enemy ships under fire and resuming the original course as soon as possible to keep guns bearing on the enemy. Individual squadron commanders had the discretion to manoeuvre their ships to avoid torpedoes, but were expected to rejoin the battle as soon as possible thereafter.[41]

Above all, the emphasis was on a full understanding of the commanding admiral's intentions, the accepted and condoned use of initiative by squadron commanders and individual captains, the understanding of the guiding principle, squadrons acting in mutual support, and a concentrated volume of fire to crush a particular point of the enemy's line of battle:

> It will be appreciated from the foregoing instructions that it is improbable that the rapid defeat of the enemy can be attained by adhering to a rigid line of battle although the signal to form line of battle is generally necessary ... The squadrons of a fleet vary in gun power and gun range, the intensity of the enemy gun and torpedo fire will also vary in different parts of the line and commanders of squadrons, both in and on the flanks of the line of battle, must take full advantage of every incident of the action to press the enemy; the action will be fought by the squadrons or divisions acting in support of and in co-operation with the squadron or division led by the Admiral, conforming generally to his movements, but only adhering to the rigid

line on which the fleet deploys when it is evidently advantageous to do so.[42]

Throughout the battle, light cruisers and destroyers would be screening the battle fleet from air attacks, as well as from any destroyer and submarine attacks that might have eluded the van forces, and would also be ready to move into their own attacking positions ahead of and on the disengaged side of the battle fleet, ready to launch their own attacks once the heavy ships were engaged. The captains of destroyer flotillas had to watch the movement of the battle fleet at all times and conform to its movement to maintain their assigned station, and not obstruct the clear observation of the enemy line of battle by funnel smoke.[43]

Obviously, the ultimate aim of the destroyers was to launch a torpedo attack on the enemy, as well as supporting the battle cruisers and cruisers breaking up any attacks made by the enemy destroyers, and the destroyer flotillas were expected to seize the first favourable moment to attack with torpedoes keeping their 'A arcs' open and allowing all guns and torpedoes to bear on an enemy. The destroyers' main roles were:

> to influence the main action; to force the enemy to turn or, if he did not, to cripple him with torpedoes. In battle, destroyers were treated as mad dogs on the leash of the destroyer flotilla commander ... Their role was to rush in a tight pack and seize a battleship's throat if they could, or, as was more likely, to leap and claw and growl at the enemy's own mad dogs which had also charged into the fray at a single word from their master ... a squadron of destroyers, bows on with a bone in its teeth was a very visible and chilling threat indeed.[44]

In reality, a massed destroyer attack on an enemy line happened infrequently, and was seldom a success. However, it remained a basic tenet that if enough destroyers fired enough torpedoes, sinkings and, just as importantly, confusion would result: 'The massed destroyer attack, launched at the correct moment when fleets are well engaged, and pressed home to the closest possible range, is a powerful and effective weapon in the hands of a British Commander in Chief.'[45] Once these attacks had been made, though, they were to resume defensive duties, stationing themselves in the van or rear of their own fleet to avoid fouling the range for the battle squadrons.[46] If enemy destroyers counter-attacked, it was the duty of the van flotillas to oppose them, whilst the rear flotillas kept on a single line of bearing to minimize the target area presented. The leader of the central flotilla had to be flexible, joining either group but taking account of both the available sea room and the strength of the counter-attack.[47]

Similarly, submarines, when accompanying the battle fleet, were expected to try to press forward on the surface and then dive once in an attacking position, targeting enemy aircraft carriers and any crippled battleships that presented themselves as targets, whilst minelayers were expected to sprint ahead of the enemy battle fleet and drop mines in their path, if possible, to disrupt their deployment.

The whole question of fleet submarines was problematical. At the end of the Great War, an attempt had been made to develop a submarine that was fast enough to keep up with the battle fleet. The result had been the innovative, but ultimate failure, *K*-class, steam-powered, fleet submarine. In the 1920s it was still hoped that diesels and electric motors would be developed enough to be able to provide sufficient power for a submarine to accompany the fleet, and they were still written into the Battle Instructions. But operating submarines near a battle fleet and in the middle of a fleet action was always going to be precarious, and the idea of them was gradually left to wither away, with the more effective patrol submarines being built.[48]

Night action had been particularly poor at Jutland, and as early as 1922 some measures were taken to rectify this, although it was never initially intended that the major surface action would be fought at night, merely that contact be maintained with the enemy fleet. This changed, however, in the 1930s, when night action was seen as a way of countering numerical inferiority. Originally, the Battle Instructions advised that if the enemy had not been defeated by nightfall, the cruisers and destroyers were to attempt to remain in touch with the enemy, and a searching force of cruisers and destroyers would be formed as well as a striking force, composed of the battle fleet and escorting destroyer screen.[49] Use of searchlights was to be kept to a minimum, and only to be trained on an enemy for the shortest possible time to assist in spotting and aiming, although light cruisers and destroyers attacking battleships, and illuminated themselves, were expected to aim their searchlights at the battleship's bridge in the hope of dazzling the bridge personnel, making their attack easier to execute.[50] Once the action was opened, using star shell to identify targets was judged superior to using searchlights as the area illuminated was greater. Echoing Cowan, as in daylight, the advantage would go to the fleet that could open fire first:

> the destructive and demoralizing effect of the first salvo of heavy gunfire may well prove decisive and the first steps taken must

invariably be governed by this primary consideration. Subsequent action must depend on the circumstances prevailing at the time.[51]

The last sentence is of particular significance, summing up, as it does, everything the Battle Instructions were designed to inculcate in subordinates. Flag Officers were expected to read the battle and force the action, not react to the enemy's moves. All ships had to have an awareness of the tactical situation and were required to keep a tactical plot, instead of relying on the flagship for instructions and inspiration; as with Nelson, before Trafalgar, the admiral's wishes were to be clearly known beforehand, and his intentions anticipated. All vessels could then cooperate in making a simultaneous, concentrated attack.

The aim of battle was the close and relentless pursuit of the enemy to complete his destruction. Whenever the action was fought, victory would only be attained by offensive action, and so, once battle was joined, it would not be broken off as long as damage could be inflicted upon the enemy. Forces should concentrate and be concentrated at a decisive point in the enemy formation, identified by aircraft, cruisers or submarines, with minimum forces elsewhere.

This was a theme that was also developed by Drax, who, in 1929, pointed out that there were three parts to a battle: *Geometry*, the movement of the ships; *Tactics*, using the weapons to best advantage; and *Morale*, or the development of the offensive spirit, the ability to take knocks in order to achieve a decisive result and the development of initiative.[52] To Drax, morale was the most important part, but the most imponderable and therefore the least studied. If all things were equal, superior tactics would win the day. But things were never equal. Even at Trafalgar, unique tactics were a subordinate factor in winning the day. Drax and Richmond believed that naval history showed that all successful British admirals had adopted a fierce, impetuous offence once battle had been joined and had been prepared to take risks. Hence the emphasis on the close-range battle, something which could only have been achieved in the recent war with a different style of leadership:

> British officers today are apt to argue, 'how can you get to close range when you start fighting at 20,000 yards?' Yet, strangely enough, in every battle in the late war we could have got to close action range and remained there, had we wished to do it. At Heligoland and Jutland, short visibility precluded long range fighting. At the Falkland Islands and on 17th November 1917, we had ample superior speed to close the enemy as much as we desired: But long range action was deliberately chosen ... If it should happen that we have to fight in waters where

mist and fog are prevalent (and it may be noted that the North Sea and Sea of Japan are conspicuous examples) it follows that the requirements of close action must receive special consideration ... The ideal, if we can attain it, is to gain some advantage at long range and then close without delay to achieve annihilation at short range.[53]

'Annihilation' was the key word. The Battle Instructions aimed to have small squadrons of ships closing with the enemy, presenting a smaller target and pouring accurate gunfire into their enemies, whilst the light cruisers and destroyers engaged with torpedoes and aircraft strafed, bombed and launched airborne torpedo attacks. This doctrine formed the basis of the tactical exercises undertaken by the Royal Navy, giving commanding officers at all levels experience in decision-making, investigating new tactical ideas and assessing possible foreign capabilities in as realistic a manner as possible.

Short of an actual war, or a study of past wars, there was no way to test out these theories, apart from wargames and exercises. The annual joint exercises between the Atlantic and Mediterranean Fleets had to reflect the realities of naval warfare as accurately as possible, and by 1929 there existed a comprehensive set of wargames rules for use in these, and other, exercises which covered the conduct of the fleet in battle.

Assessment of damage was based on three types of hits. The first, 'Vital Hits', were any hits likely to blow up the target; next were 'Speed Hits', which slowed a target down; and, finally, came 'Non-Vital Hits', the effects of which were cumulative and which would gradually reduce a ship's ability to fight. Figures for the results table were based on those of the 1922–26 exercises, reduced by 30 per cent to reflect operational degradation, including the stress of being under fire, high-speed manoeuvring, the distractions of noise, smoke and shell splashes and the unpredictability of the enemy.[54] So, for example, the chances of hitting a battleship such as the *Nelson*, *Queen Elizabeth* or *Revenge* or a cruiser such as *Kent* with a gun of 12-inch calibre or above were judged to be almost three or four times more likely than the chances of hitting a smaller target, such as a destroyer, with the same calibre at the same range.

The time taken to find the range was estimated at three minutes over 15,000 yards, two minutes for ranges between 10,000 and 15,000 yards and one minute below 10,000 yards. No allowance was made for the effects of near misses or for damage to the upperworks of a target. Consequently, as Table 4 shows, it was possible to work out the percentage possibility of a gun hitting a target of a certain size. A 15-inch gun had a 1.2 per cent chance of hitting a battleship, but only

Table 4
Rates of Hitting Table

	Battle Cruiser, Nelson	Queen Elizabeth, R-*class*, Kent	Hawkins	D-*class* Cruiser	V- *and* W-*class* Destroyer
GUN					
12-inch and above	1.2	1.0	0.8	0.6	0.3
6–8-inch	–	1.2	1.0	0.8	0.5
4–4.7-inch	–	–	2	1.6	1.0

Source: PRO, ADM 186/78, p. 37.

a 0.3 per cent chance of hitting a smaller, faster target such as a destroyer.

The method of assessing damage is primarily based on the number of Non-Vital hits which it is estimated will be required to cause 100 per cent disablement. In addition to these hits a chance factor is introduced to allow for vital and speed hits.[55] Gunfire was not the best way of sinking a ship; action damage by gunfire was essentially random, and unless it happened to be a critical hit on magazines, machinery space, screws or rudder, the only chance of sinking a battleship by gunfire alone was through cumulative damage, which was best achieved at close range. A ship could be stopped, or have its armament put out of action, and a warship that could not float, move or fight was ineffective in battle. Table 5 attempts to quantify the number of Non-Vital hits necessary to damage and eventually sink battleships. An *Iron Duke*-class battleship, or a battle cruiser such as *Tiger* or *Repulse*, would take 15 hits before being destroyed, whilst a *Queen Elizabeth* might need to be hit by 18 shells or more before it was sunk. Various other tables were carefully constructed which took into account the different calibres of gun that were firing, the size of the target, the use of aircraft to spot gunnery, the use of smokescreens, bad visibility, and enemy fire, which reduced the effectiveness of one's own firing.

When attacking heavy ships, to reflect their heavier armour protection, one 15-inch gun equalled six 8-inch guns or 12 6-inch guns, whilst for attacks on light cruisers and below, one 15-inch gun was equal to the effects of three 8-inch or six 6-inch guns.[56]

It was also important to know the optimum range for each calibre and the hit probability. It was no good, for example, expecting a heavy-calibre gun to be effective either at extreme long range, or at a short range. The Hits per Gun per Minute table, assisted with this.

Table 5
Number of Necessary Non-Vital Hits by 12-inch Gun and Above to Sink a Battleship

Hits	Early Battleships and Battle Cruisers (%)	Later Battleships and Nelson (%)
3	10	10
6	25	20
9	50	40
12	75	60
15	100	80
18		100

Source: PRO, ADM 186/78, 'Number of Necessary, Non-Vital Hits and the Percentage of Damage Inflicted on other Vessels by a 15-inch shell', p. 38.

Table 6
Effects of 15-inch Gunnery on Cruisers and Unarmoured Vessels

Hits	Heavy Cruiser (%)	Light Cruiser (%)	Destroyer (%)	Large Aircraft Carrier (%)	Small Aircraft Carrier (%)
1	20	30	100	30	40
2	50	60	–	60	100
3	80	100	–	100	–
4	100				

Source: As Table 5, p. 39.

Table 7
Effects of 6-inch Gunnery on Cruisers and Unarmoured Vessels

Hits	Heavy Cruiser (%)	Light Cruiser (%)	Destroyer (%)	Large Aircraft Carrier (%)	Small Aircraft Carrier (%)
2	5	10	50	15	20
4	10	20	100	30	40
8	20	50	–	60	75
12	40	75	–	80	100
16	60	80	–	100	–
20	80	100	–	–	–
24	100	–	–	–	–

Source: As Table 5.

Table 8
Admiralty War Game Hits per Gun per Minute Table, 1929

Gun	Target	Rounds/Gun	Range (thousands of yards)									
			15–16-inch			6–8-inch			All guns			
			Max-20	20–18	Max-18	18–16	16–14	14–12	12–10	10–8	8–6	6–4
16-inch	Queen Elizabeth	0.855	0.031	0.040	–	0.050	0.063	0.078	–	–	–	–
15-inch	Queen Elizabeth	0.985	0.030	0.036	–	0.045	0.053	0.064	0.076	0.093	–	–
8-inch	York	2.95	–	–	0.0442	0.049	0.056	0.070	0.091	0.119	0.147	–
7.5-inch	York	2.81	–	–	0.028	0.035	0.046	0.060	0.084	0.130	0.207	–
6-inch	York	3.90	–	–	0.037	0.042	0.063	0.074	0.092	0.136	0.203	0.389
6-inch Mk XII	V and W	4.50	–	–	–	–	–	0.045	0.036	0.089	0.155	0.308
6-inch Mk XXII	V and W	3.42	–	–	–	–	0.042	0.048	0.070	0.125	0.217	0.434
4.7-inch	V and W	3.9	–	–	–	–	–	–	–	0.078	0.104	0.137
4-inch	V and W	5.5	–	–	–	–	–	–	0.056	0.084	0.122	0.209

Source: Schleihauf, 'Concentrated Effort', p. 138.

The figures represent the Single Round Hit Probability, the chance that a single round had of hitting its target. These were low, but represented the realities of naval gunnery as it was at the time, derived first from theory, and then from a series of trial firings under controlled conditions.

The number of guns which could fire or bear on a target was multiplied by the rate of fire per minute, giving the potential number of hits per minute and this figure would then be multiplied by the Single Round Hit Probability to give the number of actual hits per minute. As it was reckoned to take up to 3 minutes before the range was found, this was then further multiplied by three. Probability increased as the range decreased, but what was not addressed were the variables such as the point and angle of impact, the nature and armoured protection of the structure that was hit and the penetrative and explosive characteristics of the shell.

So, if in an exercise HMS *Nelson* was firing its complete main battery of 16-inch guns at a *Queen Elizabeth*-sized enemy at a range of 21,000 yards, the calculation would be 9 × 16-inch guns, multiplied by 0.031 hits per minute (9 × 0.031 = 0.27) multiplied by 3 minutes, it being assumed that at this range it would take 3 minutes to establish the range and start hitting the target. This would give an actual result of a 0.8 chance of scoring a hit at this range.[58]

Similar tables existed for anti-aircraft fire. An 8-inch gun, firing at a rate of one round per minute, would need to fire 50 rounds to score one hit on an aircraft flying at 8,000 feet. For a 6-inch gun, firing at four rounds a minute, the figure was 15; for a 4.7-inch or 4-inch gun, firing at six rounds a minute, the figure was 13; and for a 3-inch gun, firing at eight rounds a minute, the figure was 25; the 4–4.7-inch was obviously the best heavy-calibre anti-aircraft weapon from these figures.

By comparison, the barrage fire from an eight-barrelled pom-pom gun, firing 80 rounds a minute, was reckoned to be able to achieve four hits a minute against an aircraft, with a one in two chance of achieving a hit in the first minute. The single pom-pom, firing at 100 rounds a minute had a one in 24 chance of achieving a hit after 30 seconds, and was expected to make a hit after 2 minutes, whilst the 0.5-inch machine gun could achieve one hit a minute, with a one in 32 chance of scoring a hit after 30 seconds. Multiple-barrelled pom-poms seemed, therefore, to be the best choice as a close-range anti-aircraft weapon for those ships that could mount them. Destroyers, on account of their size and the need to carry multiple torpedo tubes, would only be able to mount single pom-poms and 0.5-inch machine guns, and would rely on their speed and manoeuvrability to protect them.[59] Reliance on the anti-aircraft gun also lessened the need for a high-performance fighter to defend the fleet from attacking formations, and allowed the limited resources for naval aircraft to be used to produce aircraft which could spot, reconnoitre and drop torpedoes and bombs.

Another table helped calculate the damage a torpedo would cause, by showing the percentage of full speed a ship would still be capable of making in fine weather. A new battleship such as *Nelson* could still make between 85 and 95 per cent of its full speed after being hit by up to three torpedoes which had warheads between 350 pounds and 750 pounds. It would be reduced to around 18 knots, and would either drop behind in the battle line, or the battle line would slow, but it would still be capable of fighting and would only drop to half speed or less after six hits, with its fighting efficiency seriously impaired once it had been hit by between seven and nine torpedoes. The *Queen Elizabeth* and *R*-class battleships would only be reduced to half speed after about four or five hits, depending on the size of torpedo, with their ability to fight in the line of battle impaired after seven hits. *Hood* would only be reduced to less than half speed after six hits, and would be out of action after another three; and *Repulse* after five hits. The heavy cruisers of the *E, Hawkins* and *County* classes would be able to take two hits before having their speed and fighting efficiency reduced, aircraft carriers between three and six, depending on the ship and the size of torpedo, and light cruisers and destroyers one torpedo hit.[60]

The vexed question as to whether any battleship could resist air attack was never satisfactorily answered and, in the absence of a definitive answer, the Admiralty continued to rely on the battleship as the most important vessel in the battle fleet. The one weapon which could cause concern was the aerial-launched torpedo, and, as shown above, an attack by torpedo bombers could certainly slow down, if not sink, a battleship.

Any ship design was a compromise, and if millions of pounds were spent building and modernizing battleships, and then when war came the airmen were right and the battleships were all sunk, then that money would have been wasted. The problem was that no one knew if this was what was going to happen, despite the claims and counter-claims. If battleships were not built and modernized, and the airmen were wrong, then the Royal Navy's battle fleet would be destroyed, and the enemy's modern battleships would dominate the oceans. Because of this uncertainty, many agreed with the following:

> The British battle fleet is like the queen on the chess board; it may remain at the base but still it dominates the naval game. Properly supported by other weapons, it is the final arbiter at sea; to lose it is to

lose the game. It must not be unnecessarily risked, yet in war, without risk you can do nothing ... Our task is to sink more enemy ships with gun, torpedo or bomb than he sinks of ours.[61]

With the building of battleships forbidden until 1931, advocates of air power also felt that they had their best chance to show the superiority of air power over the battleship. And even then, the Admiralty planned to have its aircraft carrier force of six ships in commission by 1930, which would give it the largest number of aircraft carriers of any navy, operating Fairey *Flycatcher* fighters, Blackburn *Blackburn* spotters and Blackburn *Dart* torpedo bombers to minimize the threat of enemy air attack and to delay the enemy's deployment.[62]

Trials had helped to determine the accuracy of aerial weapons, and had simultaneously confirmed the limitations of air power at sea. The 1922 trials against the radio-controlled target ship *Agamemnon* showed a bombing accuracy rate of about 8 per cent at 7,500 feet rising to 10 per cent at 2,000 feet, although it was pointed out that the tests were carried out in ideal conditions against an undefended target moving at slow speed, and it was expected that battle conditions would reduce this accuracy rate considerably.[63] Similar tests, against the old battleship *Monarch*, in April 1923 concluded that the damage from a charge of 2,081 pounds of TNT hung from a boom fitted 40 feet below the waterline and 7.5 feet from the side of the ship would be minimal and the ship would still be able to manoeuvre and fight, and that it would need six direct hits with 2,000 lb armour-piercing bombs or 30 500 lb SAP (semi-armour piercing) bombs to sink or seriously disable a battleship.[64] As a result of the trials, in 1925 the Admiralty estimated that bombs below 500 lb would not seriously damage a heavy ship, and that it would take 12 such bombs to put a modern battleship out of action. These would have to be dropped from a height of at least 5,000 feet to penetrate the armoured deck, and at this height accurate bomb aiming would have been difficult; the chances of a direct hit were calculated at seven in 100 and, although speculative, the figures suggested that the latest battleships (*Nelson* and *Rodney*) were well protected.[65]

Torpedo-carrying planes promised better results and, during exercises in 1926, five hits were made from 18 torpedoes launched – a 30 per cent success rate – and when ships' avoiding action was limited by maintaining a divisional formation, five hits out of eight were obtained – a 62 per cent success rate. But because these were peacetime exercises, and as such were treated with caution, it was decided that the defence in wartime would have been effective enough to reduce this to a success rate of 11 per cent.

But even torpedo bombers would be vulnerable to anti-aircraft fire on their low-level run-in to the dropping point, and with their armoured decks and anti-torpedo bulges and batteries of quick firing anti-aircraft guns battleships were judged well able to cope with the threat of air attack.

Nevertheless, the Royal Navy did see a role for aircraft at sea, for self-defence, spotting for the big guns, reconnaissance and other duties, attacking other vessels, slowing down an enemy, and causing confusion to a deploying:

> This was symptomatic of what was to be a general Admiralty view that since (naval) aircraft could not reliably sink battleships, they would perform ancillary functions to the weapons system that could, i.e. the battle fleet. The judgment that naval aviation would help to achieve its objective more efficiently was the basis of the considerable effort devoted to it in the inter-war period.[66]

Aircraft carriers would have to operate with the battle fleet, however, and, as early as 1922, exercises in the Mediterranean with HMS *Eagle* had shown the vulnerability of aircraft carriers when they had to leave the protection of the battle fleet to turn into the wind to launch and recover aircraft. But subsequent exercises had also shown the great potential carrier-borne air attacks offered:

> The movement of a carrier depends on the direction of the surface wind, and it is generally recognized that she must neither lose touch with the Fleet nor run under the guns of the enemy, but it may not be so generally recognized how much the ability of a carrier to employ her aircraft to the best advantage, without losing touch, is influenced by the direction of deployment relative to the surface wind.[67]

This could entail the carrier losing touch with the fleet, or the aircraft not being able to see the fleet or the carrier for some time. Carriers operating in one group, as was envisaged, had to keep turning into the wind to operate aircraft, and for this reason could not easily form a part of the '*A-K*' line, as was suggested in the 1924 cruising diagram.

Exercises towards the end of the 1920s started to test out the practical application of a separate aircraft carrier squadron. It was especially important as it was felt that, when planning air attacks, *Furious*, *Glorious* and *Courageous* would all be able to range a strike of 12 torpedo bombers or 18 fighters on their flight deck at one time, *Eagle* eight torpedo bombers or 12 fighters, and *Hermes* and *Argus* six bombers or ten fighters each.[68] According to calculations and tables it would take five aircraft to be sure of scoring one torpedo hit on a battle fleet in action, and six if the fleet had the freedom to manoeuvre. If bombs were fitted instead of torpedoes, one bomb in every two would hit a battleship if dropped from 500 feet, one in eight from between 1,500 and 2,900 feet, and one in 12 from between 3,000 and 8,000 feet. Battleships would need six direct hits from 2,000 lb bombs dropped from 2,000 feet, cruisers two direct hits, light cruisers and destroyers one hit.[69] To be effective, though, an armour-piercing bomb had to be dropped from at least 8,000 feet:

> Not only is accuracy of bombs difficult to obtain from this height, but on a large percentage of days in many parts of the world cloud interference will make it impractical ... The hits obtained would probably not endanger the vitals of the ship, but the accumulated damage sustained in the larger percentage of hits to be expected with these powerful bombs would have a very disabling effect both on material and on the morale of the personnel.[70]

The more aircraft that could be included in a strike, the better, and in the absence of a suitable bomber, and with the uncertainties of obtaining good results, there was a clear preference for torpedo bombing.

It has long been felt that air efficiency on a par with either the United States Navy or the Imperial Japanese Navy was impossible as long as the Royal Navy's air arm remained under the control of the Air Ministry. During the Great War, there had been phenomenal growth in the size of the Royal Naval Air Service and the Royal Navy had been at the forefront of developing a tactical role for aircraft at sea. However, after the formation of the Royal Air Force there was no tactical connection between the work of the Fleet Air Arm and the rest of the RAF, and although the work of aircraft as a part of the battle fleet was judged to be essential, the divided control and divided loyalties inevitably caused inefficiency. For example, a shortage of personnel meant that some aircraft had to be paid off from cruisers and destroyers and the remaining aircraft had to be multi-role. But it is not true to state that between the wars gunnery dominated at the expense of air power. As well as building up the Royal Navy's aircraft carrier fleet, the Admiralty advised the Australian government to build the seaplane carrier *Albatross*, as well as cruisers, to defend trade and provide a measure of self-sufficiency in naval air power until the arrival of the Main Fleet.

It is true, though, that the Royal Navy and the RAF differed in how they saw air power, with the RAF seeing Coastal Command as a substitute for sea power, not an extension of it. And the RAF's concept of 'Unity of the Air' led to a separate 'air force', so that whereas the Royal Navy wanted to build a tactical air force capable of operating from a mobile base (aircraft carriers), the RAF concentrated on developing a strategic air force, operating from fixed bases. RAF control did cause some other developmental problems. Because of a lack of suitably qualified naval flyers the Royal Navy had to believe the experts, the RAF, when they stated that flying at sea was difficult and that aircraft deck parks and crash barriers were not practical on aircraft carriers. So the Royal Navy had to adjust the tempo of its flight deck operations and, consequently, the tactical use of aircraft at sea to suit the RAF. But it is unfair to cast the RAF merely as the villain of the piece. The RAF was aware of the type of aircraft the Royal Navy wanted. Believing that the successful interception of bombers was impossible and that the fleet would be defended by its anti-aircraft batteries, the Admiralty saw no need, for example, for a high-performance fighter, and the low-performance fighters that were produced instead had to be able to do other things as well, such as spot for the fleet's gunnery. The results in the latter part of the 1930s were aircraft such as the Fairey *Fulmar*, an escort fighter and spotter, Blackburn *Roc*, a low-performance fighter-spotter armed with a power-operated, four-gunned turret, the Blackburn *Skua*, fighter and dive bomber, and the renowned 'torpedo, strike, reconnaissance' machine, the ubiquitous Fairey *Swordfish*. Even later aircraft, such as the Fairey *Firefly*, a spotter fighter development of the *Fulmar*, and the Fairey *Barracuda*, a torpedo bomber, dive bomber and a reconnaissance machine, were still multi-role aircraft and not as successful as their US contemporaries. Apart from the *Sea Hurricane* and *Seafire*, all of the Royal Navy's high-performance aircraft in the Second World War – the *Martlet, Wildcat, Hellcat, Corsair, Tarpon* and *Avenger* – were American.[71]

But no matter how effective any of the weapons, be they shells, torpedoes or aircraft, everything hinged on the admiral in command, and on how well he had communicated his ideas and philosophy to his subordinates. Drax, for example, believed that

> Tactics in its highest form must always be an expression of the personality of a particular leader. It is therefore obviously impossible to lay down a sealed pattern plan of battle either by day or night. If the leader of any force has a dislike for close action, or a firm disbelief in night action, there is no more to be said. Those methods are not for him, and may well bring disaster to his command if he attempts them.[72]

Here, he is laying the emphasis almost entirely on the admiral, and on the fact that the right man had nothing to fear from any form of battle that promised decisive results. Even the Battle Instructions were to be guidelines only, quickly understood and executed, allowing for mobility and the effective use of all weapons. Officers joining the fleet would be able to study these as a guide to the intentions of the admiral, bearing in mind that they were not Battle Orders.

And there would be opportunities for subordinates using their initiative if a war was fought against Japan. With the Sea of Japan, like the North Sea, commonly experiencing mists and fogs, maximum visibility was likely to be low, about 12,000 yards, and an enemy fleet might have to be brought under heavy, accurate fire within a few minutes. Close action and the ability to take quick decisions had to be a feature of any tactics developed by the Royal Navy.

Increased weapon range and effectiveness made scouting all the more important, and when aircraft could not operate, cruisers would still be a vital component of the fleet. Heavy cruisers of the *Kent* class would use their high-speed and long-range 8-inch gunfire against an enemy, supporting the battle cruisers in attacking the enemy cruiser screen, whilst the light cruisers would be scouting and using their speed to give warning of an approaching enemy.[73] Disrupting the enemy's flow of information, in any way at all, gave a fleet the advantage of deploying, and possibly firing, first.

Battle cruisers would therefore drive in the enemy cruisers, aircraft would bomb the enemy aircraft carriers to deprive them of reconnaissance aircraft, fighters would strafe any personnel in exposed positions, and destroyers would launch torpedo attacks. At the same time, the fast battleships and battlecruisers would be crossing the enemy's 'T', pouring concentrated fire into his ships, at close range, whilst he was still trying to deploy, and the battle fleet would be pouring salvoes onto the enemy line.

Decisive results would only be achieved by a fleet that attacked and, in deciding upon his course of action, an admiral would need to have a clear mind as to his major tactics. Destruction of the enemy ships would be brought about by destroying the cohesion of his line of battle and the power of mutual support, and every opportunity had to be taken to achieve this.

All vessels in the fleet were expected to act in concert, with just one aim, the annihilation of the enemy, and all squadron flag officers and individual captains were expected to keep this as their priority. No weapon was to be neglected; guns, torpedoes, gas, smoke and mines were all to be used. The Battle Instructions clearly reflected the Royal Navy's integration of all weapons into a sea battle, but also the clear belief in both the quantitative and qualitative superiority of the Royal Navy. If the smaller fleet limits negotiated at Washington had imposed restraints on the Royal Navy, they had also presented opportunities to act in ways that were more in the spirit of Nelson, more in accord with Beatty's beliefs, more in tune with the Royal Navy's institutional self-image and more likely to gain the desired result – the decisive victory in battle.

NOTES

1. PRO ADM 186/66, 'Naval War Manual 1925, October 1925', p. 2.
2. Bell, 'The Royal Navy', in Siegel and Jackson (eds), *Intelligence and the International System*, p. 7.
3. J.T. Sumida and N. Lambert, 'Jutland Reconsidered', lecture, University of Hull/Society for Nautical Research Conference, 1 June 1996.
4. E. Chatfield, Admiral of the Fleet, *The Navy and Defence* (Heinemann, London, 1942), p. 114.
5. Sumida and Lambert, 'Jutland Reconsidered'.
6. NMM, Richmond Papers, RIC 11/2, 'Tactics V; Destroyers 1922'.
7. A. Gordon, *The Rules of the Game: Jutland and British Naval Command* (John Murray, London, 1996), pp. 528–9.
8. W. Schleihauf, 'A Concentrated Effort: Royal Navy Gunnery Exercises at the End of the Great War', *Warship International*, Vol. 35, No. 2, 1998, p. 131.
9. Ibid.
10. Ibid., p. 13.
11. Ibid.; W. Schleihauf, 'The Dumaresq and the Dreyer', *Warship International*, Vol. 38, No. 1, 2001, pp. 6–29, No. 2, pp. 164–201, No. 3, pp. 221–32, for a detailed analysis of British Fire Control during the period.
12. C.S. Forester, *The Ship* (Penguin, London, 1971), p. 74.
13. J. Wenzel, Director, HMS *Belfast*, to author, 31 May 1995.
14. P. Halpern (ed.), *The Keyes Papers, Vol. II, 1919–1938* (Navy Records Society, London, 1980).
15. NMM, Richmond Papers, RIC 10/2, 'Tactics', one of a series of lectures written for Royal Navy Staff College, Greenwich, 1921.
16. Churchill Archives Centre, Drax Papers, DRAX 1/18, 'Grand Fleet Battle Tactics', 16 December 1916.
17. Ibid., Drax Papers, DRAX 7/1–7/6, 'Tactics', Spring 1920, rewritten 1921.
18. Ibid., Drax Papers, DRAX 7/2, 'Tactics', 1921.
19. NMM, Richmond Papers, RIC 10/2, 'Tactics'.
20. Chatfield, *The Navy and Defence, Vol. 1*, p. 192.
21. NMM, Richmond Papers, RIC 10/2, 'Tactics'.
22. NMM, Richmond Papers, RIC 11/2, 'Tactics II', 1922.
23. NMM, Richmond Papers, RIC 11/2, 'Tactics VII: The Pursuit of Portions of the Enemy'.
24. NMM, Richmond Papers, RIC 11/2, 'Tactics III', 1922.
25. PRO, ADM 186/66, 'Naval War Manual, October 1925', p. 1.
26. Ibid., p. 2.
27. PRO, ADM 186/72, 'Battle Instructions 1922–1927, 15 October 1927', p. 1.
28. Ibid.
29. Ibid., p. 12.
30. Till, *Air Power and the Royal Navy*, p. 13.

31. Ibid., p. 7.
32. Drax Papers, DRAX 2/2, 'Battle Tactics: Lecture delivered at Malta to the Mediterranean Fleet, 1 November 1929'.
33. PRO, ADM 186/72, 'Battle Instructions', p. 6.
34. Ibid., p. 2.
35. Ibid., p. 36.
36. Ibid., p. 31.
37. Ibid., p. 3.
38. Royal Naval Museum Manuscript, 332/1994, 'Lecture Précis Notes from Senior Warrant Officers Course, Royal Naval College, Greenwich, Autumn 1925', p. 2.
39. PRO ADM 186/72, 'Battle Instructions', p. 3.
40. Ibid., p. 4.
41. Ibid., p. 5.
42. Ibid., p. 21.
43. W. Hughes, *Fleet Tactics: Theory and Practice* (Naval Institute Press, Annapolis, MD, 1986), p. 76.
44. NHB, CB3016/31, 'Progress in Tactics, 1931', 'Letter from the Commander in Chief, Mediterranean Fleet, 1931', p. 27.
45. PRO, ADM 186/72, 'Battle Instructions', p. 28.
46. NHB, CB 3016/30, 'Progress in Tactics, 1930', p. 28.
47. PRO, ADM 186/75, 'Chronological Survey of the Fighting Instructions, 1 June 1928', p. 15.
48. D. Henry, 'British Submarine Policy, 1919–1939' in B. Ranft, *Technical Change and British Naval Policy, 1860–1939* (Hodder & Stoughton, London, 1977), pp. 85–6; W.J.R. Gardner, 'Two Committees, Three Submarine Classes and 31 Hulls: the *R*, *K* and *M* classes', and E. Grove, 'British Submarine Policy in the Inter War Period, 1919–1939', both in M. Edmonds (ed.), *100 Years of 'The Trade'* (CDISS, University of Lancaster, 2001), pp. 23–9, 33–45.
49. PRO, ADM, 186/75, 'Chronological Survey', p. 46.
50. Ibid., p. 47.
51. Drax Papers, DRAX 2/2.
52. Ibid.
53. PRO, ADM 186/78, 'War Game Rules, 1929', p. 34.
54. Ibid., 'Rate of Hitting', p. 34.
55. Ibid.
56. Schleihauf, 'Concentrated Effort', p. 138.
57. Ibid.
58. PRO, ADM 186/78, 'War Game Rules, 1929', p. 43.
59. Ibid., pp. 52–3.
60. Chatfield, *It Might Happen Again, Vol. II*, pp. 100–1.
61. N. Freidman, *British Carrier Aviation: The Evolution of the Ships and their Aircraft* (Conway Maritime Press, London, 1988), pp. 90–2, 155–65.
62. G. Till, 'Air Power and the Battleship', in Ranft (ed.), *Technical Change*, p. 117.
63. Ibid., p. 118.
64. R.A. Burt, *British Battleships 1919–1939* (Arms & Armour Press, London, 1993), p. 36.
65. Schleihauf, 'Concentrated Effort', pp. 117–37. This would compare with a gunnery success rate of between 2 per cent and 3 per cent.
66. Till, 'Air Power', in Ranft, *Technical Change*, p. 112.
67. NHB, CB 3016/30, 'Progress in Tactics, 1930', 'The Tactical Employment of Aircraft Carriers, compiled by HMS *Courageous*', p. 47.

68. PRO, ADM 186/78, 'War Game Rules, 1929', p. 63.
69. Ibid., pp. 71–2.
70. NHB, CB 3016/30, 'Progress in Tactics', p. 23.
71. D. Hobbs, Curator of the Fleet Air Arm Museum, Yeovilton, to author, 21 May 2001.
72. Churchill Archives Centre, Drax Papers, DRAX 2/2, 'Battle Tactics'.
73. NHB, CB 3016/30, 'Progress in Tactics', p. 28.

6

The Royal Navy's Strategic and Tactical Exercises

Undertaking a war with Japan was recognized as being logistically difficult, given the distance and poor infrastructure, and merely having a war strategy such as War Memorandum (Eastern) was not enough to guarantee success against Japan. Nor was having a set of tactical guidelines, although these were certainly indicative of the confident attitude the Royal Navy was rediscovering between 1919 and 1939 as it tackled the problems of sending a battle fleet to the Far East and how it was to fight its enemy. Both the strategic and tactical problems needed to be refined through testing, questioning and development as an important part of what Fiske called the 'preparation strategy'.[1] Throughout the 1920s and 1930s the Royal Navy was actively doing this, setting strategical exercises based around the prospect of a war in the Far East, running tactical exercises which could evaluate the usefulness of the battle fleet and how the submarine and the aircraft carrier could be integrated into it, as well as revising the Battle Instructions. Exercises were undertaken regularly, either as staff exercises carried out at the Royal Naval Staff College, Greenwich, and the Tactical School, Portsmouth, or they could be carried out by individual vessels or squadrons, such as the annual, combined manoeuvres by the Atlantic and Mediterranean Fleets, which gave the opportunity for the fleets most likely to be sent east in a future war to practice together and refine their tactical expertise.

The problem with any exercises was in deciding their aim and what the results, or rather the lessons, were. The ideal would probably have been to have an exercise which combined strategical, tactical and training elements in one, thoroughly investigated a problem and came up with clear conclusions. In reality, for various reasons, an exercise may have concentrated on only one element, and the results would inevitably vary, depending on the level of training the participants already had; a squadron of ships which had been together for a long time, and which had previously exercised together, would perform differently to a newly formed squadron or a newly commissioned ship. They would also be dependent on what conclusions the planners wanted to test, which may have further limited their usefulness. The main object of exercises often was to allow mistakes and misjudgements to be made in peacetime, and not when it really mattered, in battle.[2] In many ways, their main value was in providing a further insight into how the Royal Navy saw itself and its role.

For planning purposes, the Admiralty had originally assumed that war was likely any time from 1929, influenced by the original guidance of the 'Ten-Year Rule' rather than Japanese intentions.[3] Among examples of the planning exercises are a series of planning problems which formed a part of a Senior Warrant Officers' course at the Royal Naval College, Greenwich, in 1925; course members were given various strategic problems to solve.[4] Whilst the aim of these exercises was more likely to have been to give course members experience in solving the sorts of problems more normally faced by strategic planners in the Admiralty, the scenarios themselves offer an interesting insight into how the Plans Division at the Admiralty may have been expecting a Far Eastern war to develop.

One exercise, Scheme X2, concerned issues of world-wide trade defence with a scenario set in US waters, following a declaration of war between Britain and Japan, with the United States staying neutral. Singapore had been relieved, but was suffering sporadic raids by Japanese battlecruisers, whilst apart from sweeps by the aircraft carriers, the Japanese battle fleet remained in home waters, waiting to draw the British battle fleet north for a decisive battle on their terms. A cruiser patrol line had stopped Japanese trade with the Dutch East Indies and Europe. Hong Kong remained in British hands and all British trade north of Saigon had ceased. Japanese naval activities were therefore mainly limited to commerce raiding by armed merchant cruisers and, in order to guard against these, a squadron had been formed to protect British trade between the United States and Australia, to cover the western approaches to the Panama Canal and to stop any Japanese trade with the United States.

Apart from the enemy chosen, Japan, there was little new in this scenario, it being very similar to the trade protection role that the Royal Navy had adopted against German raiders in 1914. Hence the scheme dealt with the disposition of the available ships following a suspected Japanese raider being sighted operating in the Caribbean, and another Japanese ship sighted passing through the Panama Canal. To counter these movements, and to maintain sea communications when not enough was known of the enemy or their possible intentions, the Staff Solution was to detach a light cruiser and armed merchant cruisers into the Pacific Ocean, placing superior forces in the threatened area.

The more useful of these exercises, from the point of view of suggesting how the Royal Navy saw War Memorandum (Eastern) working, was Scheme JI, a war with Japan. It was assumed that at some time in the 1920s Japan would take advantage of a deteriorating European situation – worsening relations between Britain and France – to press excessive demands on China, and force the Netherlands to sell Sumatra to them. As a consequence of this, the Royal Navy had sent a fleet, composed of vessels that were expected to be sent eastwards in the event of an actual war (the two *Nelson*-class ships, five *R*-class battleships, the battle cruisers *Repulse*, *Renown* and *Hood*, the aircraft carriers *Glorious*, *Furious* and *Eagle*, and a light cruiser squadron of five light cruisers, seven destroyer flotillas, 32 submarines organized into two submarine flotillas, one minelayer, minesweepers and depot and repair ships) to Singapore, where it arrived in the month of April. Costal motor torpedo boats (CMBs) for the local defence of Singapore and six submarines for the same duties at Hong Kong were also despatched; the China Squadron of five light cruisers was at Hong Kong and the light cruisers of the Royal Australian Navy were also available. In short, this was a formidable naval force, which partly reflected the Royal Navy's numerical superiority during the 1920s, and the fact that, up to 1931 at least, the Admiralty could have deployed the numbers of ships needed, although the fact that it was virtually all of the Royal Navy that was being used, and that there was no base developed at Singapore to supply such a fleet, show it to be a paper exercise.

Nevertheless, in May a force of Royal Marines, including the Mobile Naval Base Defence party, arrived, and on 5 May the Commander-in-Chief, Far Eastern Fleet, was informed that war would be declared at midnight on 14–15 May, GMT. As the situation with France had now stabilized, reinforcements were also being sent out, to arrive by August. These would be the five *Queen Elizabeth*-class battleships, four *Iron Duke* battleships, the battle cruiser *Tiger*, 12 *C*-class cruisers, three flotilla leaders and 24 destroyers and a further 20 submarines. Until their arrival, an advanced base for operations was to be established for operations against Japan's China trade.

Virtually the entire strength of the navy was being used in this strategic exercise, which was being carried out one year after the first War Memorandum (Eastern) had been written, and Scheme JI did reflect some of the contemporary thinking: the arrival of the Main

Fleet, followed by operations which would end in the decisive naval battle. Hong Kong remained in British hands, and Europe was also apparently pacified, allowing the entire Royal Navy, with the exception of three *King George V*-class battleships, three small aircraft carriers, 12 cruisers and some 60 destroyers to cope with all other naval duties.

For the purpose of the exercise, it was envisaged that, following its arrival at Singapore, the Far Eastern Fleet would sail towards the Okinawa group of islands and arrive off the island of Nagagusuku Wan on 15 May, landing the Royal Marines without opposition. These islands would certainly have been considered suitable for an advanced base for the British fleet, as they were beyond land-based aircraft range but near enough to Japan to cause problems with its trade, and they may well have been considered as a possible objective on the move north from Hong Kong.

Timings in the exercise seemed to be very optimistic, not to say unrealistic, reflecting its abstract nature. It has to be remembered that this was not detailed, strategical war planning, to be used by the fleet, but a planning exercise set in a particular context, to give course participants experience of producing answers to particular problems, and its inclusion here is as an example of how much the problem of a war against Japan was figuring in Admiralty thinking.

So, for example, by 22 May (one week after arrival, and within the timescale set for securing a base), the Mobile Naval Base Defence Organization had fortified the harbour against attacks by submarines and destroyers, and had installed the anti-aircraft defences. By 12 June, the base defences were completed, with booms and minefields, although the gun batteries were insufficient to defend the base against bombardment, and the defences were not strong enough to repel a major amphibious landing. British cruisers and submarines were operating from here against the Japanese trade to the Yangtze River and Shanghai. In the rear, Singapore, the fleet's major base, had been transformed into a protected anchorage with the storage capacity and repair facilities of Invergordon; 500,000 tons of oil fuel had arrived, and one of the ex-German 40,000-ton floating docks was positioned in the Old Strait.

The Special Idea of the exercise was that Japanese iron ore stocks were low, and so eight Japanese freighters loaded with iron ore were being prepared to run from Shanghai to Japan. Additionally, Danish and five Swedish ships loaded with oil, consigned to their own nationals not the Japanese government, had also arrived in Shanghai from Borneo, adding the presence of neutral vessels in the convoy to the problem.

Intelligence reports from an agent in Shanghai indicated that a convoy, made up of the Japanese and Swedish ships, was to sail for Karatsu (33' 29N, 129' 53E) and Wakamatsu (33' 54N, 130' 48E) on 20 June, speed 8 knots, with a close escort of six destroyers. On 21 June a squadron of four Japanese battle cruisers, one large aircraft carrier and one destroyer flotilla would arrive as distant cover for this important convoy.

The Japanese First Fleet was at Sasebo. This was made up of six battleships, three aircraft carriers, 11 light cruisers, one destroyer flotilla and one submarine flotilla and was believed to be at 2 hours' notice to steam. There were a further three light cruisers, 24 destroyers and 25 submarines available if needed. Additionally, Japanese submarines had been reported off Sidmouth Island (26' 45N, 128' 21E), Cape Yakima (26' 5N, 127' 40E) and off the entrance to Nagagusuku Wan.

Hitherto, the British aim had been to maintain the advanced base until the arrival of the reinforcements. The agent's report presented the possibility of striking a blow against the Japanese fleet, by bringing a portion of it, the convoy's screen, to action. The whole of the British fleet, less the battleships *Ramillies* and *Royal Sovereign* (involved in a collision with each other), were available, and although the Japanese were judged to be slightly superior in fighting power, the British battle cruisers had a 4-knot superiority in speed over the Japanese vessels. Sasebo lay 440 miles away, 22 hours at 20 knots. The rendezvous point for the Japanese convoy and covering force was 330 miles away from Sasebo, 16½ hours at 20 knots, and 440 miles away from the British base at Nagagusuku Wan, 22 hours at 20 knots.

It was judged to be unlikely that the Japanese would attack the base with the convoy operation in progress, but more likely that they would sortie with the First Fleet, to support the covering force and convoy. It seemed to be in British interests to attack the battle cruiser force before the Japanese battle fleet could intervene, and not seek an action with the Japanese battle fleet before the arrival of the reinforcements. The position of the Japanese battle cruisers was known, and they were an inferior force, hampered by their responsibilities to the convoy. Defeating them would increase the British force's numerical superiority, and adversely affect Japanese morale. In undertaking this mission, the risks of accidentally meeting with the entire Japanese battle fleet, and of leaving the advanced base undefended, had to be accepted. The scene was being set for a 'classic' action, a raid against a convoy, designed to defeat a weaker portion of an enemy fleet and draw out the main enemy strength in its defence.

The British battle fleet could, it was believed, leave Nagagusuku Wan undetected by aircraft as the base lay beyond the range of air reconnaissance, and destroyer patrols could keep Japanese submarines down whilst the fleet left, the other known patrol lines off Sidmouth Island and Cape Yakima being avoided. Leaving harbour after dark, and observing wireless silence at sea, would further hamper detection.

Always assuming that they would not be subjected to the same counter-measures, British submarine patrols off Sasebo could give warning of any Japanese sorties, and air patrols from the carriers at sea would give timely warning of the approach of any enemy force. Employing the whole force should ensure that the Japanese battle cruisers would be enveloped, cut off from Sasebo and destroyed.

Scheme JI, in particular, offered insights into what contemporary Royal Navy thinking may have been and suggested how the Royal Navy's operational planning may have been developing for a Far Eastern war. The British battle fleet had moved, for the purposes of the exercise, virtually unopposed and without loss, to a point within striking distance of Japan, acting against trade and a detachment of the Japanese fleet, which had come out to protect a convoy. Scheme JI was just one of a whole series of exercises and, given its form, not unnaturally lacked realism. But it is no less useful as an example of contemporary strategic thinking for all of that, and a later exercise in the Mediterranean Sea, Exercise MZ, presented a similar scenario, this time using the ships of the Atlantic and Mediterranean Fleets.

In many ways these operational and tactical exercises, which could involve a few ships, a squadron or a fleet, were the most useful. They trained crews in their duties, allowed squadrons to work together and allowed flag officers to get to know their squadrons' strengths and weaknesses. A former Admiral of the Fleet, Sir Edward Ashmore, recounting his time as a midshipman on *Rodney*, the flagship of the Home Fleet, commented:

> My chief recollections are of endless gunnery practices and sizeable fleet manoeuvres at sea. At one time we had the 2nd Cruiser squadron, some destroyer flotillas, the carriers *Courageous* and *Furious*, and the R class battleships *Ramillies*, *Royal Oak*, *Revenge* and *Resolution* at sea in company, with the battlewagons usually astern of the Fleet flagship in line ahead.[5]

Ashmore is, perhaps, giving the widespread, but limited view. Although the emphasis was undoubtedly on surface actions, a lot of different exercises did take place, in fact, and all were useful in resolving strategical and tactical issues. For example, until the completion of the Singapore naval base, the passage and operations of the Main Fleet in the Far East rested on the provision of oil reserves along the route and the use of anchorages secured by the Mobile Naval Base Defence Organization (MNBDO), itself still in an embryonic state. The planned conversion of the battleship *Agincourt* to an MNBDO depot ship had been halted by the Washington Treaty, and the Royal Marine formations necessary for a fully functioning MNBDO were still not authorized. Nevertheless, Exercise EA, held in January 1922, had tested the concept of escorting a slow convoy, including a MNBDO ship, when the Atlantic Fleet sailed from Portland to Arosa Bay. A Red Force of seven battleships, the First Light Cruiser squadron and two flotillas of destroyers escorted the battle cruiser *Courageous*, along with the repair ships *Assistance* and *Sandwich* and the depot ship *Pandora*. The aim was to relieve the depot ship *Maidstone* and the sloop *Snapdragon*, representing the MNBDO ships, protected by mines, submarines and a local defence flotilla, whilst opposed by a Blue Fleet of two battle cruisers, the Second Light Cruiser Squadron, two destroyer flotillas and submarines, the sort of force with which the Japanese were expected to raid the Indian Ocean. A similar exercise, Exercise NASF, was held in Greek waters in 1926 to test the feasibility of a full-scale deployment of a MNBDO.[6]

A further example was Exercise MU, which took place between 15 and 17 August 1925 in the Mediterranean, as one of the combined Atlantic Fleet and Mediterranean Fleet exercises.[7] The underlying objective was to ascertain the likely risks to which a British Main Fleet would be exposed when passing through the southern part of the Malacca Strait on its way to relieve Singapore (heavily invested on land by the Japanese) to the north, with the Japanese battle fleet operating from a base in the Dutch East Indies (near Singapore), intent on crippling the British fleet on its passage through the narrow waters of the Strait.

A location in the southern Aegean Sea was chosen for the exercise, the route from Cape Matapan to the Doro Channel via the Elaphonisos Channel. Argostoli represented the secret anchorage at Nancowry, Singapore was represented by Doro Island, One Fathom Bank by the narrow neck of Elaphonisos Channel, Port Swettenham by Vatika Bay, with the Japanese base represented by Vitali Bay.[8] The exercise was designed to replicate as accurately as possible the passage through the narrow, mined Malacca Straits, and forces were allocated as far as possible in the proportion that might be expected in the event of a war. As there was only one aircraft carrier and one flotilla of submarines available, these were both allotted to the Blue, or Japanese, force, as their presence was deemed to be vital to any Blue plan. No mention was made of Japan, or the area under consideration, in fleet orders, although flag and commanding officers were informed of the real nature of the problem. This was both to keep the nature of the exercise secret and to fend off any prying queries from the Foreign Office, who had instructed the Admiralty not to prepare for a war against Japan.

In his 'Remarks By Commander-in-Chief', Admiral Keyes evaluated various aspects of the exercise. He did not feel that any major conclusions about the problems of relieving Singapore could be drawn, although he pointed out that paravanes and high-speed minesweeping gear had both worked well, which led him to comment that it would be better to attempt to approach Singapore by day and to risk any Japanese minefields, as any losses would cause less disorganization and sweeping should be more effective. He wrote:

> On the whole it is considered that crossing by day is preferable so long as a strong force of fighters can be put into the air to prevent the fleet being subjected to torpedo plane attack while in the minefield ... It has been clearly shewn that in an operation of this nature the fleet could not afford to wait while deliberate sweeping operations took place.
>
> The consequent necessity for the full development and fitting to destroyers of the Oropesa sweep were emphasised. During the exercise RED was forced to use as a clearing sweep, a sweep that is only designed as a searching sweep.[9]

He also noted that the submarine attacks were well carried out and that

> The air work generally was good, and the reports accurate. The tactics in the first torpedo plane attack were very successful. The formation made use of the clouds to conceal their approach and when over the battle squadrons at 4,000 fleet broke formation and attacked individually. The attack developed from all directions with the result that up until the last moment, they were thought to be reconnaissance aeroplanes and the imminence of the attack was not realized.[10]

The battleships *Queen Elizabeth*, *Barham* and *Marlborough* were all judged to have been hit and slowed down by air-launched torpedoes. These were specific and important, tactical points, which needed to be known.

It was recognized that the relief operation would be a hazardous undertaking, sailing through a narrow channel which could be heavily mined and contested by enemy light forces. It was vital, therefore, that Singapore was developed as a base able not only to maintain a large fleet, but also to defend itself until the fleet could arrive and drive off any attackers, as well as to support and repair the fleet in subsequent operations.[11]

Other exercises investigated other problems of naval warfare, and whilst they were mainly to test the efficiency of ships and squadrons in a range of scenarios, many would be applicable to a war in the Far East. The results of these, published annually, sometimes seem obvious, repeating what had been learned by hard experience during the Great War, and the real value of these exercises was as training, allowing captains and flag officers to exercise their command and develop their use of initiative. For example, between 1925 and 1927, exercises investigated the problems of escorting a large, slow troop convoy, exposed to attack by battle cruisers and light cruisers. As the vulnerability of the convoys was recognized, the strongest escort possible was required, and the importance of close shadowing at night and careful sifting of reports were also highlighted.[12]

The use of air reconnaissance was seen to be a mixed blessing, unless special tactics were evolved for the aircraft carriers. Whilst aircraft had extended the scouting distance by up to 150 miles, the possibility existed that not only was it easier to find the enemy fleet, but that it was easier to be detected by the enemy's aircraft as well, or that a returning aircraft would be shadowed by an enemy aircraft. And, as the aircraft carriers needed to be with the battle fleet for protection, once the aircraft carriers had been spotted, so, too, would the battle fleet. Exercises in the 1930s continually showed that aircraft carriers which operated away from the battle fleet were vulnerable to surface and submarine attack, and, in a war with the Japanese, especially in the approach to Singapore, the loss of the aircraft carriers would be a disaster. Recognizing this, the solutions adopted by the Royal Navy were to operate the aircraft carriers with the battle fleet, and to develop the armoured aircraft carrier and rely on the massed, anti-aircraft barrage for protection from air attack. The Japanese, by comparison, whilst also recognizing the vulnerability of aircraft carriers, developed in a different way, and emphasized the conversion of suitable mercantile hulls as a way of producing a large number of aircraft carriers quickly and cheaply. They also opted to operate their aircraft carriers as a separate squadron, eventually with cruisers and battleships attached as protection.

Other exercises confirmed the importance of combined air and surface reconnaissance, as, for example, Exercises MK and BD, when light cruisers, destroyers and aircraft located, shadowed and attacked a battle squadron. Exercise MQ showed up the difficulties in handling more than three destroyer flotillas at the same time in a night attack – an important tactical lesson, if destroyer attacks were to be successful against a Japanese fleet. Exercises also showed up the difficulty of protecting up to 40 merchantmen from attacks by a determined enemy, something that would need to be considered if a large troop convoy was approaching Singapore or Hong Kong in the presence of the Japanese. Unless control of the sea was secure, the passage of a troop convoy would not be justified.[13]

These exercises were primarily tactical and training exercises, to maintain a level of proficiency and to build up professional knowledge. All of the exercises showed an awareness of some of the various problems that naval warfare in the 1920s presented, and how the Royal Navy was attempting to counter them through training and the use of improved tactical doctrine. However, none of the conclusions drawn seems remarkable (issues such as the need for escorting troop convoys or using searchlights), while others would never really be resolved until the actual test of war, as the Royal Navy would find out to its cost in 1939 when using fleet aircraft carriers in an anti-submarine role.

The problems associated with the successful relief of Singapore were readdressed in March 1928, with Exercise MU2, 'The Relief of Alboran Island', in which over 80 ships of the Mediterranean and Atlantic Fleets tested the idea of securing a victory through a decisive action, and which was judged to be one of the most impressive and realistic of all the exercises carried out to test the problems associated with the relief of Singapore.[14]

This was the fleet that would form the major elements of the Main Fleet in time of war, and the exercise was obviously a rehearsal for the relief of Singapore. Alboran Island, off the Algerian coast, represented Singapore, and Gibraltar represented Hong Kong, closely invested by the Japanese. A Red (British) defended base existed on Alboran Island, the area around which was defended by shore batteries up to a range of 8 miles. The base, under attack for some time, was in imminent danger of falling. Blue had no land-based aircraft and was operating at a considerable distance from its bases, but had almost complete control of the seas. The fall of Red's base would jeopardize all future operations against Blue, and its speedy relief was therefore of the utmost importance.[15]

A Red force – the Mediterranean Fleet – of ten battleships, four battle cruisers, five cruisers, two destroyer flotillas and one aircraft carrier formed the Main Fleet, steaming to the island's relief. This was opposed by a Blue force of six battleships, four battle cruisers, eight cruisers, six destroyer flotillas, one aircraft carrier, one minelayer and 13 submarines. The Alboran Channel could only be entered from the east, and as it had been announced by Blue that no neutral shipping was allowed in it, it was presumed to be mined.

The object of the exercise was, again, to examine the comparative merits of passing through a narrow strait, approaching the entrance in the dark and approaching the exit point in daylight, and the alternative, passing into the entrance in daylight and exiting in the dark, in an area in the vicinity of which the enemy had been operating for some time.

Each side had a Special Idea, an Appreciation of the Situation and a Proposed Course of Action. The Blue Special Idea obviously was to capture the island before the arrival of Red and if this appeared doubtful to fight a fleet action against Red, aiming to reduce the Red Fleet's effectiveness:

> Our object has hitherto been the reduction of ALBORAN ISLAND in conjunction with military and air forces. The near proximity of Red Fleet, the probability that Red Fleet will proceed direct to the relief of ALBORAN and the existing state of defence of the Island requires a change in our object.
>
> To relieve ALBORAN ISLAND, Red Fleet must of necessity pass through the ALBORAN CHANNEL which fixes him to our great tactical advantage. By making full use of our advantageous tactical position we should be able so to reduce the strength of Red Fleet as to enable our main fleet to deny its passage to ALBORAN ISLAND.[16]

In other words, the most vulnerable part of the assault was again seen as the final approach, through contested waters, and it was necessary to determine when the best time would be to attempt this passage.

The Blue Appreciation of the Situation confirmed Red had a superiority in battleships but not in light forces, and from a tactical standpoint the superior speed of four of Red's battleships would be of no use in the Alboran Straits. Blue therefore had to aim to inflict the maximum amount of damage whilst the Red Fleet was passing through the Straits and had limited manoeuvring room, exploiting Blue's superiority in light craft and avoiding a daylight surface action until mines and torpedo attacks by destroyers and submarines had reduced Red forces. This really decided the Blue course of action: to lay mines as far to the east of the channel entrance as possible, and to launch airborne torpedo attacks whilst the Red ships were in the minefield. These attacks would be followed up by torpedo attacks from up to four destroyer flotillas. Submarines would also attack, and, if Red were sighted in the channel in daylight, torpedo planes would continue their attacks. If Red chose to have an advance force of destroyers to sweep the mines, supported by cruisers and battle cruisers, this would also favour Blue, who would use the advantage in light forces to launch more torpedo attacks on this advanced force.

The Red Appreciation of the Situation recognized that Blue was working at a considerable distance from home ports and had no shore air bases. Red expected to be opposed, but did not expect Blue attacks until the Red Fleet was deploying in the Alboran Channel, as to do otherwise would be to throw away the principal advantage of an excellent defensive position.[17] Attacks by mines, submarines and destroyers were all anticipated once the Red Fleet was in the channel, and it was also acknowledged that proceeding straight up the channel would expose the fleet unnecessarily to attack, and any such attacks might delay the passage of the fleet, exposing the fleet to attack by light forces.

Regarding the principal object of the exercise, whether to sail up the channel in daylight or at night, it was decided that the Red Fleet's arrival at the entrance of the channel by day would allow the Blue forces to keep Red under observation, and attempting the passage at night entailed great risks from destroyer attacks. What was eventually decided upon, therefore, was not to proceed straight up the channel, but to accept the risk of Alboran falling, delay the arrival of the fleet for 12 hours, and wait until daylight and a time when the tides would be high enough to allow the fleet to pass over the minefields. The fleet would then pass up the Alboran Channel, preceded by the destroyers streaming sweeping gear, then by the cruisers in line ahead as cover, then the battle fleet and aircraft carriers. Air reconnaissance would have been flown off early, and if the Blue aircraft carriers or battleships were sighted, these would be attacked. Otherwise, reconnaissance and anti-submarine work would be priorities for aircraft. A passage in daylight would also maximize the advantage Red had in battleships, ease the task of the aircraft carriers and force Blue into a battle. The Red planners believed that if Blue left engaging until darkness fell, then there was every possibility that Red would slip past the Blue battle fleet and reach Alboran.

A second solution had also been provided for Red as, in a departure from the norm, the Staff College at Greenwich was given the opportunity to consider and offer comments before the exercise took place, rather than afterwards. The members of the Staff College concluded:

> Under the conditions of the Exercise it seems likely that the Red Main Fleet will enter the Channel between the hours of 0300 and 0500 if he elects to come in during the first night or between midnight and 0500 on subsequent nights.[18]

The admiral of the Red fleet had decided on the same course of action, approaching at dawn to gain the maximum advantage from daylight, minimizing the risk of mines, and allowing for the use of aircraft to support the fleets and reduce the danger of a surprise torpedo attack by Japanese forces. And, as advised by the Battle Instructions, the confusion of a night action was avoided and all of the ships of the fleet could engage the enemy.

A later exercise, carried out in July, showed the importance attached to naval aviation, when the aircraft carriers *Eagle* and *Courageous* were on opposite sides in a Mediterranean Fleet exercise. Torpedo bombers, escorted by fighters from *Courageous*, made an attack on *Eagle* and found it a 'sitting duck', with aircraft still ranged on its decks. It presented the ultimate target for any carrier air group, vulnerable and defenceless. Unsurprisingly, it was judged to be destroyed.

The problems associated with bringing a portion of the enemy fleet to battle were explored in Exercise MZ, held in March 1929.[19] In a situation similar to that of Scheme JI four years earlier, MZ was an attempt by a weaker battle fleet to intercept and destroy a portion of a superior enemy fleet, detached from the main fleet for convoy protection, in effect, forcing the enemy (the Japanese) into battle.

The strategical game was again conducted at Greenwich, six days before the actual exercise, on 12 March 1929.[20] A Red Force of eight battleships, three battle cruisers, seven or eight cruisers, two aircraft carriers, four destroyer flotillas and six submarines was at sea, screening two convoys. Red and Blue had been at war for some years and the situation was now reaching a critical point. The Red Fleet had achieved a measure of superiority over Blue, whose fleet had been reduced to making sporadic raids on Red territory and trade routes. Because of these raids, a division of four Red battleships regularly supported the convoy escorts, occasionally joined by cruisers and aircraft carriers.

Owing to repairs and refits on Red battleships, a temporary equality had been established between the fleets, and the Blue government wanted their fleet to attack a convoy and make a successful demonstration of strength, restoring morale and the will to fight among the population. As it was not known which convoys would be escorted, and because the loss of battleships was seen as a blow to the morale of the enemy's fleet, if the Red battleships were met, they, and not the convoy, would become the main target.

One Red convoy sailed at 18.00 hrs on 17 March, followed by a convoy of troop transports the next day, and one division of battleships sailed to support both convoys. The rest of the Red Fleet was at short notice to sail, as the Red Admiral was aware of the likelihood of a raid.

Air reconnaissances flew up to 145 miles from their carriers and, in consequence, the Blue Fleet was soon sighted, and the main portion of the Red Fleet sailed. The convoys were dispersed to neutral territorial waters, to preserve them as much as possible, and the two portions of the Red fleet sailed to combine.

As soon as the Red convoy and covering force had been sighted, two flights of Blue torpedo bombers were launched, whilst the Blue Admiral pressed on to intercept the covering force with his fleet, unaware that the rest of the Red Fleet was at sea. The Blue battle cruisers engaged the Red covering force at extreme range, whilst airborne torpedo attacks continued, damaging one Red battleship, although a Blue battle cruiser was also damaged in attacks by six Red torpedo bombers. Red forces withheld further air strikes, however, hoping to draw Blue further away from their main base. Blue's own air reconnaissance forces had meanwhile located the main Red fleet, steaming for the action, and as neither of the Blue aims, the destruction of the convoys or the destruction of the Red covering force, could be achieved, the Blue Admiral concentrated all of his forces and fell back, illustrating, perhaps, that it was unrealistic to expect every contact to result in an action.

The performance of the aircraft, flying up to 145 miles from their carriers, played a key role in this exercise. Equally significant was the number of wireless signals made. Afterwards, the vice admiral commanding the Blue 1st Battle Squadron wrote at length about the use of wireless to control the fleet:

> I feel a word of caution about W/T facilities is necessary. Since the advent of short wave W/T, restrictions on W/T communications have been considerably reduced.
>
> We place so much reliance on being able to signal to all our forces that we do not look ahead as far as we used to and are apt to give orders or instructions to Senior Officers of units without knowing what the situation is.
>
> While fully realizing the great value of the facilities we enjoy I do say that they should be used with due discretion or we may arrive at centralized control instead of centralized direction. This may result in checking the initiative of subordinate commanders producing a system of waiting for orders.[21]

This is an extremely telling point, highlighting as it does the need to remind commanders about the use of wireless, and also because of the implied distinction between 'centralized control' and the more flexible 'centralized direction' that the new Battle Instructions were advising. The eradication of the culture of presuming that the admiral saw and knew everything, apparent at Jutland, was one of the most important tactical features of the period, and the widespread use of short-wave wireless posed a potential threat to this which, if not addressed, could have negated any progress that had been made. There was obviously the opportunity to use wireless as another means of signalling, along with searchlights and flags, but the danger in so doing was that it may have had the effect of stopping an admiral from having a long-term battle plan and, instead, just reacting to enemy moves, thus acting defensively. Similarly, squadron commanders could have been tempted to stop using their own initiative, and merely waiting for and relying on signals from the fleet flagship. In the context of a war against Japan, it is again indicative of a desire for subordinates to use their initiative within the boundaries of the Battle Instructions in a decisive battle, and not wait for orders.

Aircraft were becoming increasingly important in these exercises. A warship is useful only if it can move, float and fight. Remove one of these three conditions and it becomes useless. The only way to disable a warship was to either blow it up and sink it, or put enough holes in it to make it sink, or else damage its capacity to move, fight and repair damage. The quickest and, in many ways, the most effective way of doing this was by disabling or sinking the vessel with aerial torpedoes, allowing the battleships to catch the disabled enemy fleet and engage with gunfire. But operating aircraft meant working a battle fleet with aircraft carriers, which led to having to screen and protect the aircraft carriers, tasks which slowed down the battle fleet, as aircraft carriers had to turn into the wind to launch and recover their aircraft.

The solution was a simple alteration to the accepted cruising formation. Some 20 years before the war in the Pacific, the Royal Navy was facing the problem of operating aircraft carriers with a battle fleet and offering the solution of a separate aircraft carrier squadron. By the end of the decade, in Exercise MI on 23 March 1929, the aircraft carriers *Furious*, *Glorious* and *Courageous* were operating as a squadron, the first such in any navy, able to range strikes of up to 12 torpedo bombers or 18 fighters on deck at one time, compared to the smaller strikes of eight torpedo bombers or 12 fighters from HMS *Eagle*, and six bombers or ten fighters from either *Hermes* or *Argus*.[22]

Not unnaturally, this exercise highlighted problems as well as successes, the main one being that a mix of aircraft in each ship was better than a dedicated air group. One carrier had been the reconnaissance carrier, one the spotter and fleet defence carrier, and the third the attack carrier, but the exercise showed that the loss of one ship then denied the fleet all of one type of aircraft. A further exercise confirmed that, with an adequate escort, the aircraft carrier could act as the centrepiece of a fast striking force. The same exercise also showed up the relative merits of torpedo bombing and level bombing. To be effective, bombs had to be dropped from a height of 8,000 feet, and this made accuracy difficult and would not necessarily cause serious damage to a ship unless several hits could be obtained.

More was expected of the torpedo bombers, with their potential to blow holes in ships and disable or sink them, and a squadron of Blackburn *Ripon* torpedo bombers launched an attack on the enemy fleet at an unprecedented distance of 110 miles, flying to the target and back without the aid of a special navigational aircraft, a real achievement for the aircrew and aircraft of the time. In the *Confidential Admiralty Monthly Intelligence Report* for September 1929 the role of the naval aircraft was confirmed and the report stated: 'A slogan (which is now becoming well known) is that "there are three F's in naval warfare – FIND, FIX and FIGHT". Aircraft do the first two and the fleet, assisted by its aircraft, does the third.'[23] In the following year, a study, *The Tactical Employment of Aircraft Carriers*, was compiled by the air staff of HMS *Courageous*. Using the experience of the exercises, this suggested an independent role for the aircraft carriers, freeing them from the battle fleet:

> It is therefore put forward that under favourable air conditions the air squadron, supported by battlecruisers, one squadron of cruisers and two flotillas of destroyers, might form a separate unit stationed well ahead of the Fleet. The Senior Officer of carriers should be given complete freedom of movement.
> The advantages are:–
>
> (i) The carriers are not hampered in their movements as it is always easy to fall back on the advancing fleet wherever the wind. The whole detachment conforms to the movement of the carriers.
> (ii) Early torpedo attacks can be launched.
> (iii) The carriers have protection from surface light craft and submarines.[24]

Admiral Sir Reginald Henderson had been appointed by the First Sea Lord, Sir Frederick Field, to be Rear Admiral Aircraft Carriers in 1931, with command of *Courageous*, *Furious* and *Glorious* in the Mediterranean, where he operated these ships as a group, developing dive-bombing techniques, coordinated torpedo bomber attacks, massed strikes and long-range escorted missions, rather than the traditional roles. Operating aircraft carriers in this way was also something that both the United States Navy and the Imperial Japanese Navy were evaluating. Whilst they took it on and developed it, the comparative lack of expertise and aircraft caused the Royal Navy merely to refine its existing naval aviation policies, supporting the battle fleet.

In many ways, one of the most notable inter-war fleet exercises was carried out by Admiral W.W. Fisher, Chatfield's second in command in the Mediterranean and the Commander-in-Chief of the Mediterranean Fleet from 1932, and it is often used as an example of how far the Royal Navy had developed its tactics since the Battle of Jutland. Like Chatfield, Fisher believed in the effectiveness of night fighting and continued to develop his fleet's proficiency in such tactics, and his handling of the Mediterranean Fleet in the naval manoeuvres of 1934 – its interception of the Home Fleet (Blue) and a 'convoy' of troop ships at Arosa Bay on the coast of Spain after sailing his fleet through a gale and engaging his unsuspecting enemy at full speed, at a range of 7,000 yards, just after midnight – ended the idea that the Royal Navy could not fight an enemy at night.

The Red Fleet (the Mediterranean Fleet) of five battleships, seven cruisers, two aircraft carriers, 27 destroyers and five submarines was opposed by a Blue Fleet (Home Fleet) of five battleships, two battle cruisers, one aircraft carrier, eight cruisers, 19 destroyers and four submarines. Blue was tasked with raiding Red trade and also establishing a base on the Spanish Atlantic coast (*Eastland*), whilst the Red Fleet's task was to intercept this invasion fleet.

Even with the intelligence as to when Blue left the Azores, Red would not know precisely where the Blue fleet was heading for. Admiral Fisher reasoned that there were only two likely spots for a Blue base, Arosa Bay in Spain or Lisbon in Portugal. Both were well placed to attack trade, but Lisbon was only 250 miles from Gibraltar and was easily blockaded. He determined on heading for Arosa Bay. In fact this was precisely where the Blue Fleet was heading, planning to travel the last 100 miles at night to escape detection.

A heavy gale blew up shortly after the Blue Fleet left the Azores, making air reconnaissance impossible for either side to take advantage of. Fisher decided to cover Arosa Bay and Blue's possible line of advance with his destroyers, supported by his battleships, steaming 70 miles eastwards of them.

Blue's battleships and destroyers were to be 50 miles to the southeast of the transports, whilst the battle cruisers, an aircraft carrier and some cruisers were to be to the south of this to mislead the Red Fleet as to Blue's true intentions. But Blue's own destroyers, which were at Lagos Bay, Portugal, were unable to put to sea during the gale, depriving the Blue admiral of a second element of his force. Nevertheless, he decided to proceed with his plan.

The Blue battle cruisers were sighted by Red's destroyers, and shortly afterwards the battle fleet was also sighted. Fisher ordered his destroyers to move northwards, and his fleet effectively split the Blue Fleet in two, holding a central position. Towards evening, in heavy weather, the Red battle fleet moved towards the unsuspecting Blue battle fleet and engaged them at a range of 7,000 yards. The exercise was declared closed, with a victory to the Red Fleet. Their success was due to many varied factors. Their destroyers were more modern, and more seaworthy than the First World War vintage Blue destroyers, seamanship was superior, Fisher had made good strategic judgements, taking risks and closing with his battleships to 7,000 yards, whilst the Blue admiral had been hamstrung by the political directive to establish a base. In a real wartime situation the Blue operation probably would have been cancelled, but the exercise did show that British ships could fight at night, in foul weather. Rear Admiral Cunningham, flag officer of the Mediterranean Fleet's destroyers, wrote:

> These bold and masterly tactics not only put an end to the exercise; but settled once and for all the much-debated question as to whether or not British heavy ships could and should engage in night action against corresponding enemy units. At Jutland, it will be remembered, night action had deliberately been avoided. From now on it became an accepted principle that a highly trained and well-handled fleet had nothing to lose and much to gain by fighting at night.[25]

Fisher was hailed by the press as a hero and called 'the greatest naval genius of modern times'. Not quite everyone agreed, and according to Admiral Sir Roger Bacon, writing in 1942, Fisher had had more than his share of luck. Bacon, rather out of time in his ideas, believed that it should never be the intention to fight the decisive fleet action at night, because

> The admiral commanding the stronger and better practiced fleet will never engage in a night action if it is possible for him to avoid doing so. The reason for this is that any night action is a pure gamble. The battle range is reduced to some 3,000 yards, a point blank range, so all gunnery efficiency and refinements are thrown to the winds, and chance and the torpedo brought into play ... Of course, if the enemy will escape unless fought at night, then fight you must, but it is far better if possible to wait until daylight before joining action.

> Destroyer attack is bound to complicate a night action and be an aid to the less practiced fleet; since all close quarter fighting favours the more inefficient side, by depriving his adversary of the advantage of the skill that long practice and discipline have created in the vessels which are under his command.[26]

If the enemy had not been defeated by nightfall, night actions were to be avoided if at all possible and, instead, Bacon advocated forming a Searching Force of cruisers and destroyers to maintain contact with the enemy until the following morning, when the battle fleet could re-engage, exactly what had *not* happened at Jutland. Bacon was writing about battle from another era, and Madden's Battle Instructions had been revised to reflect this. Even if the Royal Navy did not intend to fight its decisive battle at night, it could not afford to ignore this as a possibility and, Bacon's views notwithstanding, night fighting became a feature of all British tactical training and an integral feature of the Battle Instructions. It was yet another way for a fleet inferior in numbers to ensure a speedy victory. And there is no doubt that night fighting, if properly practised and exercised, can be effective. Cunningham's performance at Matapan bears this out. Another writer, Admiral Sir Studholme Brownrigg, emphasized the importance of quick reactions, with or without orders, when, also in 1942, he wrote that 'The watchword [of night action] has been dubbed "impomo", that is, the "instantaneous production of maximum output".'[27] Certainly, by the outbreak of war in 1939 the Royal Navy had embraced the idea of fighting at night, recognizing that British naval supremacy could no longer be maintained by numbers alone, and British tactics needed to change to accommodate the possibility if fighting an enemy superior in numbers and in ships with the advantage of long-range gunfire. Only *Nelson* and *Rodney* were able to engage ships at a range over 30,000 yards, unlike the battleships of the Japanese (and those of the German) navy. During a daylight battle, unless the British ships had the advantage of speed, that is, unless the torpedo bombers or destroyers had slowed down the enemy line, then the enemy ships would be able to keep the range open. Exercises in 1933 and 1934 had confirmed the difficulties that the Royal Navy would face in trying to close the range during daylight, and night action seemed to offer the better chance of success.[28] Night action could not only allow a Royal Navy battle fleet to maintain contact with an enemy fleet, but also give the advantage of darkness to the destroyers. Night action also favoured closer ranges and Brownrigg's 'impomo', and regular night-fighting exercises were held with ranges of 8,000 to 9,000 yards, the targets being illuminated by cruisers.[29]

In fact, it was becoming increasingly apparent that the only chance that British forces would have against the Japanese battle fleet would be to seek an engagement at night or when visibility was poor, and pass as quickly as possible through the danger zone *en route* to Singapore. For example, if a *Queen Elizabeth*-class battleship was engaging a *Fuso*-class ship, the ideal range was estimated to be 12,000 yards to 15,000 yards, whilst the *Queen Elizabeth* was deemed to be most at risk between 17,000 yards and 19,500 yards. During daylight, the Japanese ship would have been able to manoeuvre to keep the range open but at night the close ranges would secure the greatest chance of hitting the target, as well as coinciding with the British 'zone of immunity'. At longer ranges hits were likely to be fewer and a waste of precious ammunition. In fact,

> The Service's willingness to accept a night action was an attempt to mitigate through tactics its increasing strategic naval inferiority. This inferiority only worsened as ... ships could not be built to the dimensions of their rivals because of the limitations in port facilities and the constraints presented by the Suez Canal ... Given the restricted displacements of the second *King George V* and *Lion* classes of battleships and their limited ammunition loads of 80 and 60 rounds per gun respectively, an engagement fought at long range was no longer a tactical option for the Royal Navy.[30]

The Royal Navy, in common with other navies, was constantly exploring potential problems, and conducting various tactical exercises to determine the responses to these. Although many of the conclusions reached remained dependent on specific results from specific sets of circumstances, undoubtedly the main value of all of these exercises was as training, allowing officers and men to make 'high order' mistakes where they could be analysed and corrected, away from the stress of battle. The Royal Navy with its strategic war plan, War Memorandum (Eastern), investigated whether Singapore could hold until the arrival of the Main Fleet, whilst the tactical exercises considered the problems the fleet might encounter in relieving Singapore, as well as differing tactical situations, designed to bring out the best from individual ships and squadrons, and increase familiarity with the Battle Instructions.

The conclusion reached by the Admiralty in the first decade after the Great War was that, with the right conditions, the right ships in the right numbers and with the right training, Singapore could be relieved, and a naval war against Japan successfully undertaken. From 1931 onwards, with changing conditions in Europe and an increasingly ageing battle fleet which needed replacing or modernizing, a more cautious attitude began to prevail and the strategic planning began to reflect this, even though the tactical planning reflected the Royal Navy's confidence in its ability to 'do the job'.

In reality, the only way to successfully test out a strategical or tactical development is to have it operate in time of war, and no matter how good operational plans may have been, they rarely survived intact after contact with an enemy, so perhaps the real value of these exercises lies in the fact that they were presenting the opportunities to develop a high level of training and expertise, within the framework of the most likely scenario, a war with Japan.

The Royal Navy would have been able to despatch 15 battleships to the Far East if a war broke out before 1931 and, tactically, the question was how best to use this fleet. Strategic success depended on more than just superior numbers and if a war was to be fought against Japan, even with the existence of the naval base at Singapore, operations would need to be conducted at great distances from dockyard support. Operations would need to be meticulously built up and battles resolved quickly, and the Royal Navy's tactics were developed with this aim in mind. The Naval War Manual of 1925, for example, clearly stated that battle was to be regarded as the decisive factor in naval war and the destruction or neutralization of the enemy's battle fleet the aim and object of the Royal Navy. Without this, control of sea communications could not be guaranteed, and so all training was to be directed towards this.[31]

Strategically, both the Naval War Manual and War Memorandum (Eastern) stressed the main principles of war: the objective, offensive actions, surprise, concentration of force, economy of force, mobility and cooperation. Tactically, and not unlike Madden's Battle Instructions, the Naval War Manual stressed the need for concentrating the battle fleet whilst in action until it was clear that the enemy had been defeated. Concentrated firing would break up the cohesion of the enemy line, disable the enemy battleships and gradually allow for a greater and greater concentration on the remaining enemy battleships. As sea battles were more than just a straightforward engagement between battle fleets, tactical training, initiative and a thorough knowledge of doctrine were to form the basis of tactical training and exercises during the 1920s. Nothing could compensate for the courage, determination and aggressiveness that was expected of the modern commanders, and these qualities would only be developed through continued exercising. Even as late as 1935, Drax was making the point that having manuals of orders and instructions would have the psychological effect of the men relying on them in battle, and success would be more assured by a clear understanding of the spirit of the admiral's plans and an urgent desire to take the initiative, so that when the unexpected happened a good officer should react and make a decision, not turn to the relevant manual. The Battle Instructions reflected not only the integration of all weapons into a modern sea battle, but also a clear belief in the quantitative and qualitative superiority of the Royal Navy. A future naval battle would have aircraft bombing and strafing the enemy, destroyers darting forward to launch torpedo attacks and battleships closing to close range to overwhelm and defeat their opponents. Development of the qualitative features of naval warfare could only come through wartime experience or, in peacetime, an analysis of wartime experiences, exercises and training. What differed from Jellicoe's Grand Fleet Battle Orders was the emphasis on initiative, the assumption that the commander-in-chief knew everything that was going on and that subordinate squadron commanders should wait for orders before acting. The repeated testing of theories and tactical situations at fleet, squadron and ship level emphasized the importance of using initiative and that naval warfare was more than a question of numbers of ships, but was a synthesis of the quantitive, tradition and training. Exercises were vital, as ships, officers and men unaccustomed to constant, rigorous training would be hesitant in battle. Exercises therefore had to be as realistic as possible so that any mistakes would be made in training, and not in the actual battle. Consequently, throughout the 1920s and 1930s all aspects of battle fleet operations were tested and refined, divisional tactics were explored, night-fighting skills developed, communications improved and aircraft integrated into battle tactics, all with the aim of improving the Royal Navy's ability to fight and decisively win any future battle.

But the Battle Instructions also made it clear that whilst subordinates were not to expect a succession of signals or instructions from the flagship once battle was joined, the admiral would be still in overall control, and squadron commanders had not only to observe the flagship and follow its lead, but also be familiar with the admiral's intentions in battle. The many annual publications concerning exercises and tactical developments, some of which have been referred to in this book, were the way of keeping abreast of developments, they were not manuals to be referred to on the bridge.

The Naval Staff in 1928 did not feel that night fighting was desirable, and felt that it could lead to great confusion. But it was a way of evening up the advantage and well-handled, well-trained crews would be capable of handling such tactics. Under Chatfield's command between 1930 and 1932, night-fighting tactics were regularly practised in the Mediterranean Fleet in an attempt to inculcate the principles of strategical and tactical surprise, exactly what Fisher had achieved at Arosa Bay. Fighting at night would also, of course, make the use of aircraft by both sides less effective. The Royal Navy remained in ignorance of Japanese expertise in night fighting, however.

The development of divisional tactics was also a move away from the tactics of Jutland and towards a system which depended on judgement and initiative. Brought about by the necessity of using a smaller battle fleet – the Mediterranean and Atlantic Fleets were usually only five or six battleships strong – battleships were now to be fought in divisions of from three to five ships, with more scope to manoeuvre independently, and supporting each other through concentrated firing. Another advantage was that a smaller battle fleet presented a smaller target, especially to massed destroyer attacks. The enemy destroyers would need to get closer and pick targets, rather than just launch spreads of torpedoes, to be sure of hitting a battleship. The same would be true of an air attack, when ships could close up if necessary and support each other with an anti-aircraft barrage.

When Drax started to rewrite the War Plans in 1935, he reiterated the ideas he had first put forward in the previous decade. Namely, that in battle complete success would only come from an understanding of the admiral's plans and an urgent desire to take the offensive. Officers of all ranks were expected to develop the mentality which would allow them to take the initiative, so that when the unexpected happened in a naval battle, they would deal with it. He wrote as much in a letter to Admiral Backhouse in September 1935, pointing out that the type of battle as written down in the Battle Instructions was that least likely to happen. In his lecture notes on initiative, Drax gave examples of what he meant *should not* happen. In a PZ exercise in the Mediterranean, an 8-inch gun cruiser had joined the fleet on the disengaged side, opposite the centre of its own line of battle, and stayed there until it was ordered to engage the enemy. Similarly, during night exercises in 1931 a group of destroyers following astern of some heavy ships observed enemy ships passing on the opposite course and, instead of immediately attacking, their leader signalled for permission to attack. By the time it had been given by the flagship, the targets had gone. In a subsequent lecture, Drax also quoted from the Official Report of the 1931 Fleet Exercises:

> From the report of the battleship *Nonsuch* – 'When the *Blank*, (Blue cruiser), passed down the line to starboard at 2137 she was seen by *Nonsuch*, (Red battleship). It was assumed that other units of the battle

fleet had also seen her, and since no action had been taken, this was interpreted as being the intention of the Senior Officer.

This cruiser then remained stationed on the starboard quarter at a range of about 5,000 yards until 2159 during which time she was, presumably, shadowing the battle fleet and reporting them.

The question which called for an immediate decision in this case was whether by opening fire on this cruiser and sinking her, more definitive information would have been conveyed to the attacking force as to the position of the battle fleet than she was able to supply herself by making enemy reports unmolested.

The likelihood of her being engaged and driven off by *Douglas* who was astern, was taken into account and no action was taken against this cruiser.[32]

The reply, from the Flag Officer commanding the Red Fleet, was:

> The Blue cruiser sighted by *Nonsuch* was not observed in Flagship. Once inside effective range this cruiser should undoubtedly have been engaged. Hostile light forces should never be allowed to shadow if they can be sunk or driven off. They are potential fighting units, if not a present torpedo menace. Moreover, the reports they pass are certain to be of value to the enemy. These considerations far outweigh those which prompt withholding fire. A gun action, seen from a distance, does not necessarily provide the enemy with more reliable information than is already possessed. It may even mislead. The necessity for sinking or disabling enemy units which are within effective gun range is even greater when evasion is of importance than when it is relatively unimportant.[33]

Jutland was still casting a long shadow in some quarters, but it was only by constantly practising, by making the errors such as these, in peacetime, that the Royal Navy was ever going to become efficient enough to fight a war at sea against a determined enemy.

NOTES

1. Fiske, *The Navy as a Fighting Machine*, p. 142.
2. Naval Historical Branch, London, W.J.R. Gardner, to author, 26 August 1999.
3. J. Moretz, *The Royal Navy*, p. 165.
4. Royal Naval Museum Manuscript 332/1994, 'Senior Warrant Officers Course, Royal Naval College, Greenwich, Autumn 1925'.
5. E. Grove (ed.), *The Battle and the Breeze: The Naval Reminiscences of Admiral of the Fleet Sir Edward Ashmore* (Alan Sutton, Stroud, 1997), p. 12.
6. Moretz, *The Royal Navy*, p. 166.

7. P. Halpern (ed.), *The Keyes Papers, Vol. II, 1919–1938*.
8. Ibid., p. 152.
9. Ibid., p. 154.
10. Ibid., pp. 153–4.
11. Roskill, *British Naval Policy between the Wars*, Vol. I, p. 538.
12. PRO, ADM 186/43, 'Selected Reports of Exercises, Operations and Torpedo Practices in HM Fleet, Summer and Autumn 1927; Summary of Exercises, Spring 1925–1927'.
13. PRO, ADM 186/43, 'Selected Reports of Exercises, Operations and Torpedo Practices in HM Fleet, Summer and Autumn 1927, Volume II', p. 63.
14. PRO, ADM 203/86, 'Exercise MU2; The Relief of Alboran Island, 15 March 1928'.
15. Roskill, *British Naval Policy*, p. 538.
16. PRO, ADM 203/86, 'Blue Special Solution', p. 2.
17. Ibid.
18. Ibid., 'Letter from The Director, Royal Naval Staff College, Greenwich, to the Chief of Staff, Atlantic Fleet', p. 1.
19. PRO, ADM 186/145, 'Exercises and Operations 1929, Vol. I'.
20. PRO, ADM 203/90, 'Strategical Exercise MZ, Atlantic and Mediterranean Fleets, 12 March 1929'.
21. PRO, ADM 186/145, 'Exercises and Operations 1929', p. 83.
22. Ibid.
23. Moretz, *The Royal Navy*, p. 231; ADM 223/817, 'Confidential Admiralty Monthly Intellegence Report, No. 124, 15 September 1929'.
24. NHB, CB3016/30, 'The Tactical Employment of Aircraft Carriers', compiled by HMS *Courageous*, in *Progress in Tactics, 1930*.
25. Admiral Viscount Cunningham of Hyndhope, *A Sailor's Odyssey: The Autobiography of Admiral of the Fleet Viscount Cunningham of Hyndhope* (Hutchinson, London, 1951), p. 161.
26. Admiral R. Bacon, 'Fleet Battle Tactics; Night Actions', in Bacon (ed.), *Britain's Glorious Navy* (Odhams Press, London, 1941), p. 267.
27. Admiral Sir Studholme Brownrigg, 'Gunnery: Night Actions', in Bacon (ed.), *Britain's Glorious Navy*, p. 221.
28. Moretz, *The Royal Navy*, mentions Exercise 'RR' and 'ZJ' in 1934, p. 226.
29. Ibid., Exercises 'OX' and 'PC' by the Mediterranean Fleet in October 1930 and August 1931 respectively, and Exercise 'BA' by the Atlantic Fleet in March 1931.
30. Ibid., p. 30.
31. PRO, ADM 186/66, 'Naval War Manual 1925'.
32. Churchill Archive Centre, Drax Papers, DRAX 2/2, 'Tactics 1935'.
33. Ibid.

7

Japanese Naval Strategy and Tactics in the Far East

There were close similarities between the strategic and tactical thinking of the Royal Navy and that of the Imperial Japanese Navy during this period, most probably due in no small part to the fact that both the Imperial Japanese Navy and the Royal Navy had developed close links during the first 20 years of the century. British shipyards had built many of the first Japanese warships, the Royal Navy had trained many of the Japanese officers and men, and the Imperial Japanese Navy had adopted many of the Royal Navy's customs and traditions, including a tradition of victory, the victory over the Russian Fleet at Tsushima.

Between 1912 and 1915, Japan had constructed three of the world's most powerful battle cruisers and subsequently designed a further six battleships of even greater firepower, so that, at the end of the Great War, Japan was the possessor of the world's third-largest navy.[1] And in 1919 Jellicoe identified Japan as Britain's major potential enemy, an indication of the growing disquiet with which the Japanese naval expansion was held within the Royal Navy.[2]

British fears about Japanese expansion did not end there, however. In 1922 the Admiralty received worrying intelligence of a proposed pact between Sun Yat-sen, the Nationalist leader in China, and a Japanese company which would have allowed the Japanese to take over control of the Paracel Islands off the south coast of China. Not only would this be an infringement of the Washington Nine Power Treaty, but it would also be a threat to the security of Singapore and the route to Hong Kong. The fears were unfounded for the scheme never materialized, but the incident added fuel to the Admiralty's belief that, sooner or later, Japan would want complete regional and naval superiority. However, many British politicians remained sceptical about the idea of a war between Britain and Japan breaking out, and dismissed any idea of a Japanese threat.

Japan's major, national concerns in the period up to 1939 were essentially the same as any other nation's: national security and access to raw materials and export markets. Japan was an industrial nation, capable of producing its own ships, aircraft and a range of other industrial goods, but its industrialization had been tremendously expensive, both financially and politically. And with few colonies, such as the British possessed in the region, Japan was totally dependent on exports and imports from other countries. In 1920 it ceased to be self-sufficient in food and many Japanese economists and politicians began to feel that their country had to gain control of some alternative and secure sources of raw materials abroad, to end the dependence on foreign trade. In 1922 the Prime Minister, Korekizo Takahashi, pointed out that the recent war had shown Japan that national expansion through force of arms could not succeed and that economic competition was the only way that a nation could increase its wealth, power and standard of living.[3] But if this was so, and if this was going to lead to the British and Americans trying to stop future Japanese economic growth, then perhaps the only way of guaranteeing a supply of necessary raw materials was to exert a more direct control over China, risking war by challenging both the United States and its policy of an 'open door' concerning trade with China, and Great Britain's widespread, commercial interests in the region. Japan would then have to be strong enough to exert its influence, by force if necessary, and prevent other powers from threatening the lines of communication and trade with China.

Pursuing such a policy made Japan vulnerable. In the *British Naval Staff History of the Second World War in the Pacific*, the Japanese problem was clearly and succinctly put.[4] Japan was a country with a large population, few natural resources and a considerable industrial complex, dependent on its ability to import and export, and thus vulnerable to economic downturn and, in case of war, economic blockade. Japan's only saving grace would be that whilst an economic blockade would undoubtedly affect Japan, it would be slow to take effect, giving the Japanese military and naval forces time to inflict serious setbacks on their enemy's forces. Japan's economy could not support a long war but could inflict a tremendous one-off blow.

Japan's long-term aim, therefore, was to make the Japanese economy as self-sufficient in raw materials as quickly as possible and to defend these raw materials by drawing the enemy into a decisive battle, at a time and a place the Imperial Japanese Navy chose. In order to achieve this, since 1907 the Japanese had developed the twin operational doctrines of attrition and interception of an enemy fleet. Japanese strategists, like their British counterparts, had concluded that the best way of removing threats to Japan's trade was by annihilating the enemy's naval power and, like the British and Americans, saw that the best way of achieving this was by drawing the enemy battle fleet towards Japan, sinking as many ships as possible *en route*, before a climatic surface naval battle. Plans were developed primarily against the United States, and only in the 1930s were they expanded to include the possibility of a war against Britain, either alone or as a US ally. Lieutenant Commander Ishimaru Tota's book, *Japan Must Fight Britain*, translated into English in 1936, claimed that Japan could confidently defeat the British, and warned that war could only be avoided if Britain stayed in the southern Pacific Ocean, around Australasia, and left the western Pacific, including China and Malaya, to Japan. However, it was not taken too seriously at the time, and was dismissed by some in Japan as ridiculous and regarded with some amusement by the British.[5]

In essence, the broad Japanese strategy was very similar in outlook to both War Memorandum (Eastern) and the Americans' War Plan Orange. In the event of a war with Britain, the first phase probably would involve an attack on Hong Kong, aiming to capture the colony and defeat, or at least neutralize, the cruisers of the China Squadron, making it impossible for them to act against Japanese trade or troop convoys. Singapore would be attacked, by strong raiding forces, whilst other forces acted against Malaya and possibly the Netherlands East Indies. Meanwhile, Japanese battle cruisers and armed merchant raiders would have penetrated the Indian Ocean to prey upon the trade from Australia and New Zealand.

Phase two of the Japanese plan would involve consolidating and strengthening their newly won positions, forming a defence perimeter that the British would have to fight through before they could start a serious offensive against Japan's trade routes and possessions across the Pacific. Trade with the United States was a particularly vital element of the Japanese economy, and the British, in fact, could only intercept this near the United States, thus risking diplomatic confrontations over the rights of neutral vessels, or nearer Japan, which would expose British warships to the constant risk of raids and attacks by Japanese air and surface units, as well as submarines. In the meantime, the Japanese would exploit the natural resources in their conquered areas, simultaneously using appeals to nationalism, trying to turn the indigenous populations against the old colonial powers.

Phase three, the attrition phase, would see Japanese forces seeking out the enemy forces threatening their perimeter defences and vital areas, and attacking them. This would entail continued Japanese raids in the Indian Ocean, carrier-borne raids on Singapore and Hong

Kong, and small-scale naval engagements. The purpose of these actions would have been to reduce the total fighting strength of the Royal Navy as it attempted to implement its own war plans. The main Japanese aim was to preserve their own battle fleet, whilst drawing the British forces further into range of their land-based air forces. At the right moment, the Japanese would move to the interception phase and launch a massive land, and sea-based counter-attack, designed to eliminate the British battle fleet as a potent fighting force. With the main British battle fleet neutralized or defeated outright, British cruisers would lack the heavy support necessary to attack Japanese trade, and would be unable to oppose Japanese attacks on British trade. Negotiation or surrender would be the only options.

In order to implement these plans successfully the Japanese originally felt that it was vital to possess a navy composed of the most modern battleships and battle cruisers, all heavily armed and under six years old, and supported by the older vessels of the navy, the 8:8 Plan. This reasoning was based on a commonly held maxim of naval warfare, that an attacking fleet needed a 50 per cent superiority in firepower over a defending fleet and that the defending fleet needed to be at least 70 per cent of the strength of an attacking fleet to give it any chance of obtaining superiority in battle. It was an attractive, and simple to understand, maxim:

> the 70 percent ratio presented several advantages. It was so simple as to amount to almost a slogan; it sounded reasonable in that Japan sought to build only 70 percent of the strength possessed by the United States; and for some politicians, if not the admirals, it obscured the limits of Japan's own industrial base which could probably not build a navy larger than 70 percent of the US Navy, in any event.[6]

At the time of the Washington Conference the Japanese Delegation had expected difficult negotiations, but no one expected Secretary of the Navy Hughes' dramatic cuts as a way of limiting naval expansion. The Japanese seemingly did have a lot to lose. In January 1921, Japan had a battleship, four light cruisers and three destroyers completing and four battleships, four battle cruisers, six light cruisers and three destroyers, plus numerous minor warships building or projected, with another 55 ships, including two more battleships and four more battle cruisers, projected in order to complete the 8:8 Plan. The total cost of this was estimated to be a third of Japan's total national expenditure, and taxes needed to be high to achieve this. The Imperial Japanese Navy was committed to this plan as the minimum necessary for the defence of Japan and Japanese interests, and suggestions in the

Diet in February and March for some sort of limitation of the programme had already been rejected. Amidst continued press suspicion and speculation that the Washington Conference was a ploy to weaken Japan's ability to defend itself, the Japanese delegation attended the conference committed to the 8:8 Plan, and would only agree to reduce to a 70 per cent ratio with the other navies if the other major powers reduced their rearmament programmes, which would include not developing any bases in the western Pacific.

Japanese delegates somewhat reluctantly abandoned their 70 per cent figure and accepted a 5:5:3 ratio, or 60 per cent, in battleships, combined with an agreement not to develop any more fortified bases in the western Pacific, giving the Japanese regional security and superiority as neither the United States nor Britain could easily deploy a battle fleet to the region and neither side had a major base nearer than 3,000 miles away, Pearl Harbor for the Americans and Singapore for the British.

As in Britain, Japanese newspaper reactions during the Conference negotiations were mixed. The *Hochi Shimban*, for example, felt that Japan was in a cleft stick, as accepting the offer would make the Imperial Japanese Navy inferior, whilst if Japan refused the offer of arms limitation they would be branded as militaristic. Another paper, *Nichi Nichi*, stated that as long as the United States was expanding its navy, then so should Japan, whilst the editor of *Kokumin*, reflecting the suspicions some felt towards British motives, felt that the proposed naval holiday was a way to stop the United States overtaking Britain as the largest naval power. Japan, in signing the Treaty, was being dragged into this dispute whilst simultaneously not being allowed to build up its navy.[7]

In March 1921, the *Japan Advertiser*, an English language newspaper, ran a series of articles which analysed the 8:8 Plan and the possible effects of Washington on it. The completion of the Plan would make the Imperial Japanese Navy larger than the Royal Navy and second only to the United States Navy, a source of national pride, whilst the Washington proposals would leave Japan with a battle fleet of ten ships, well below the 16 ships the navy felt necessary for the defence of Japan. In these circumstances, it seemed to make sense to complete the 8:8 Plan first and then discuss naval arms limitation, as this would then give the Imperial Japanese Navy 16 capital ships less than 8 years old, compared to the United States Navy's 18 and the Royal Navy's nine. Similarly, Japan would have 30 cruisers and 120 destroyers, as compared to the United States' ten cruisers and 309 destroyers and the Royal Navy's 70 cruisers and 212 destroyers. But, inevitably, such a course of action would not only strain Japan's economic resources, it would probably also start a new naval race as each power attempted not to be left behind the others' new construction programmes.

The chief spokesman for the opposition to the Washington Treaty in Japan was Admiral Kato Kanji, who pointed out that the sticking point for the Japanese was the tonnage allocations, threatening as they did both the 8:8 Plan and the necessity of a 70 per cent ratio in total naval tonnage.[8] Japan's only possible courses of action had been either to gamble that the Americans would not engage in a naval race as it would cost them too much and the British could not afford to build more ships, leaving the Japanese to complete the 8:8 Plan and then negotiate from a position of strength, or to go along with the rest of international opinion and accept the Washington ratios. Japanese delegates chose the latter course of action and agreed to a 60 per cent ratio. But even maintaining a fleet at this strength caused financial problems, and by 1929, when Japan was starting to feel the effects of the international economic depression, the issue of naval expenditure came into the public eye again.

Although the 8:8 Plan itself did not survive the Washington Conference, the reasoning behind it, that if enough enemy battleships could be sunk or disabled on the approaches to Japan then the Imperial Japanese Navy, carefully concentrated, could overwhelm their opponents, survived in subsequent Japanese strategic thinking. And following the Washington and London Naval Treaties, Japan developed weapons, strategies and tactics to compensate for the limitations imposed on its war potential, to enable its navy to operate the twin doctrines of interception and attrition.[9] Heavy cruisers, fast enough to act with the destroyers and heavily armed enough to act as the fast division of the battle fleet, were built to replace the cancelled battle cruisers, whilst the appearance of aircraft and submarines added to the existing arsenal of guns, mines and torpedoes as a way of reducing their enemy's forces.

Another factor that the Japanese saw as very much working in their favour was the very size of the theatre of operations. The United States Navy would have to deploy from Pearl Harbor in Hawaii and the Royal Navy from Singapore, both then working their way towards Japan over a distance of 3,000 miles. The key to defeating an enemy would therefore be to reduce the strength of the enemy battle line through attrition attacks by aircraft, torpedo attacks by submarines and destroyers, minefields and engagements between light forces, before the decisive battle. Initially, the Japanese planned for this somewhere in the vicinity of the Ryukuku Islands, although as aircraft became increasingly effective a battle further away from Japanese territory was envisaged. If, after the Washington Conference, the Japanese could not out-build either of their rivals and were limited in what they could build, by numbers, type and tonnage, then the strategy of interception–attrition, coupled with a qualitative approach to shipbuilding and design, building faster and more heavily armed ships, would work very much in the Imperial Japanese Navy's favour.

As in US and British naval circles, the period between 1922 and 1924 was thus one of readjustment for the Japanese navy, as ten battleships were retained and 14, ranging from old, pre-dreadnoughts, to ships still at the planning stage, were scrapped, disposed of or converted to training ships. Building was kept down to what some in Japanese naval circles regarded as a bare minimum, but was actually a reasonable amount of construction. Between 1925 and 1930 the Japanese laid down two aircraft carriers (converted from the hulls of the *Kaga* and *Akagi*), four heavy cruisers, seven light cruisers, 30 destroyers, 25 submarines, three minelayers and two gunboats.[10] The strength of the Imperial Japanese Navy in September 1930 was thus six battleships, four battle cruisers, four aircraft carriers (including the still-to-be-completed *Ryujo*), eight heavy and 21 light cruisers, 104 destroyers, 67 submarines, nine coastal defence battleships, 13 gunboats, five minelayers and ten minesweepers, and around 30 other minor vessels, a far from inconsiderable fleet and one which was superior to the fleet that either the United States Navy or the Royal Navy could hope to have on station in the event of war. And with only British and American light cruisers normally based in the region, Japan had definite regional superiority.[11]

As with the Royal Navy, but for different reasons, numbers and types of cruisers became an issue for the Japanese as they, and not battleships, became the measure of naval strength. At the Geneva Conference in 1925, the expected clash was between the United States and Japan over the latter's insistence of a 70 per cent ratio of cruisers. The Japanese heavy cruiser design sprang from the Navy General Staff's quest for an all-purpose warship, one that could retain the traditional reconnaissance function of the cruiser, substitute for the battle cruiser in the decisive fleet encounter and which also had the fire power to blast through enemy screens and allow the Japanese destroyers to deliver a crushing torpedo attack.[12]

Britain, with its 15 'Treaty' cruisers, wanted to limit the Japanese to 12, which it had already authorized. But the United States was pressing for 23 such ships, and if it went ahead with these, Japan would want to build at least four more, and Britain, whose strategic planners did not favour more heavy cruisers, preferring to have a greater number of modern, light cruisers, would be forced to follow suit in order not to be left behind. However, if the United States could be persuaded to reduce its total to 21, Japan would need only two more cruisers. Better still, if it could be persuaded to build only 18, Japan would not need to build any more such ships, and neither would the British, who could then concentrate on building light cruisers. But the United States wanted the Japanese to accept a 60 per cent ratio for heavy cruisers, and would not accept the British demands for more light cruisers for trade protection. The conference collapsed in deadlock.

And during 1928, in order to relieve some of the problems caused by the Depression, the Land and Business Profit taxes were given to local prefectures in Japan, which had the obvious effect of reducing the amount available for national expenditure. The effects of this on the navy were reported to London by the Tokyo naval attaché; there was no money to refit the *Kongo* or *Hiei*, or to provide sufficient planes for the 1st Aircraft Carrier Squadron, resulting, according to a Japanese newspaper, *Hochi Shimban*, in the laying up of the aircraft carrier *Kaga*. It was also reported in the same newspaper article that when the eight 10,000-ton cruisers were completed in 1931–32, all of the dockyards would have to be temporarily shut down. In Kure, a naval dockyard town, the local edition of *Nichi Nichi* for 6 January reported a similar story: if another naval conference limited capital ship construction further, all the dockyards would have to close.[13]

The similarities with the Admiralty's predicament are striking. As in Britain, the Japanese naval leaders were anxious to keep construction going, and to keep the specialized skills of the dockyards available. A total of 25 million yen was needed for new construction in 1931–32, with another 75 million yen in 1932–33 and 100 million yen in 1934–35, but the money provided in 1929 was some 60–65 million yen, well under the desired amount for the Naval Estimates. As with the Admiralty and the Treasury, the Japanese Naval Staff argued in the Imperial Diet that the money was vital if the fleet was to be kept to the Washington limits, but the government line was to keep the fleet at its present size until circumstances changed, as a result of either further naval conferences or a deteriorating international situation. Despite this, the Naval Staff intended to introduce to the Diet a Bill calling for the construction of a fleet of nine capital ships, five cruisers, 47 destroyers, 32 large submarines and six minor warships, to be started in 1931–32. The battleships would cost 90 million yen each and the total cost of all of the ships was estimated at some 530 million yen, or 170–180 million yen a year. It was also intended to increase the number of naval air squadrons from eight to 17. Unsurprisingly, the Japanese Treasury could not approve such a programme and opinion in naval circles was that the fulfilment of the naval programme was impossible. Foreign building was watched with interest and probably also with some trepidation.[14]

The next conference to discuss naval arms limitations, the London Naval Conference of 1930, saw Japanese newspaper reports commenting on what Japan's position should be. Since Washington it had modernized four battle cruisers, built eight heavy cruisers, ten light cruisers, three aircraft carriers, 42 destroyers, 40 submarines and various auxiliary vessels and had under construction a further four heavy cruisers, 12 destroyers, six submarines and numerous minor vessels. According to several editorials, it was going to be vital that Japan did not lose out again, and in the Osaka *Manichi*, of 31 July 1929, an attempt was made to explain why.

The article pointed out that Great Britain and the United States would probably want some sort of postponement or limitation on 'Washington' type heavy cruisers armed with 8-inch guns, the very ships that Japan saw as vital to support its line of battle. For the United States this would mean delaying or cancelling three cruisers, leaving it with 20 heavy cruisers. Great Britain would suspend two such ships, but would still have built 19 cruisers carrying 7.5-inch or 8-inch guns (five *Kent* class, four *London* class plus *Australia* and *Canberra* for the Royal Australian Navy, two *Norfolk* class, four *Hawkins* class and two *York* class). Japan, by comparison, had completed the four *Furutaka* class and was building eight 10,000-ton cruisers, of which two, *Nachi* and *Myoko*, were completed. This left six vessels under construction, which stood to be cancelled if such a limitation was agreed, leaving Japan distinctly inferior, with only six such cruisers. According to the English edition of the Tokyo *Nichi Nichi*, Japan was going to press for a 10:10:7 ratio for cruisers, which would give it the scope to construct another six to eight vessels, whilst the United States and Britain would be unable to build any more.[15]

At the London Naval Conference Japan agreed to limit itself to 108,400 tons of Class 'A', or heavy, 8-inch gunned cruisers, 100,450 tons of Class 'B', or 6-inch gunned light cruisers; and 57,000 tons of submarines. However, concerns about the suspected US naval programme, which reportedly included modernizing ten battleships, building heavy cruisers and building an extra 1,000 naval aircraft, led Japan to consider some sort of response and the Japanese started to develop their naval air force as a striking force to replace the shortfall in the number of cruisers. Japanese naval aviation doctrine started to move away from scouting and reconnaissance towards attacking enemy fleet units, and techniques for successful horizontal, dive and torpedo bombing were all developed. A strong, land-based air contingent was also established to support the fleet, and tactics developed to utilize these diverse arms in a coordinated manner, one that would support the interception–attrition strategy and that would be utilized in the final, decisive battle.¹⁶

On 7 October 1930, Navy Minister Abo wrote to the Japanese Prime Minister on the subject of Japan's naval strength, stating that Japan had to launch an immediate, retaliatory programme of naval construction if the gap that existed between the strength of the two navies was not to widen. Japan's response was to deliberately set about building, not just up to the specified maximum tonnage limits for each class of vessel but beyond, exceeding the limits for tonnage and armament if necessary, in an attempt to improve the fighting qualities of its vessels. The First Naval Armament Replenishment Plan, also known as Plan 1, received Diet approval in 1931, and it aimed to construct 39 vessels (four *Mogami*-class cruisers, six *Hatsuharu*- and six *Shiratsuza*-class destroyers, seven large, and two medium submarines and minor vessels), totalling some 72,905 tons, by 1936. By the end of 1933, 25 of these vessels had been completed, although the completion date of many was delayed following the need for the Japanese naval dockyards to reduce top weight on many warships, following the widespread typhoon damage to Japanese naval craft in 1935.¹⁷

The Imperial Japanese Navy was already investing heavily in naval air power and also had asked for 28 air units to be formed, although this was subsequently reduced to 16, of which 14 would be available by 1936. These units would be two carrier-borne fighter squadrons of 16 aircraft each, six carrier-borne bomber squadrons, also composed of 16 aircraft in each squadron, four squadrons of six flying boats, one and a half squadrons of reconnaissance seaplanes for use on cruisers and battleships and half a squadron of training aircraft, marking a massive increase in Japanese naval aviation.

When Japan's relations with the international community deteriorated sharply following its invasion of Manchuria in 1931 and subsequent withdrawal from the League of Nations, the Navy General Staff exploited the situation and played up fears of a US-led, armed solution to the situation. They pressed for a Second Naval Armament Replenishment Plan, Plan 2, to be established before the previous plan had even been completed. Plan 2 was to run from 1934 to 1937 and would supply another 87 ships. Like Plan 1, the lack of dockyard space caused several ships to be delayed by between nine and 16 months.¹⁸ But the completion of the *Ryujo* added a fourth aircraft carrier to the fleet and allowed more importance to be given to the naval air arm, whilst giving the Japanese further expertise in using aircraft at sea. Battleships and cruisers routinely carried aircraft for spotting and reconnaissance and this, along with the increasingly important land-based naval air force, helped secure the aircraft as a vital component in Japan's overall strategy to defeat an adversary. The only drawback was the formation of the new air units. Eight had been planned, to bring the planned total up to 39 by the end of 1936. However, only four of these eight new units had been completed by the end of the year, delaying the number of aircraft that could actually be deployed at sea and, hence, the realism of Japanese naval exercises with aircraft.

Following the abrogation of the Washington Treaty and their refusal to agree to further limits on battleships, the Japanese commenced a third Replenishment Plan, which aimed to give the navy a further 66 ships and 14 air groups:

> Every effort was to be made to compensate for lack of quantity in naval strength by the improvement of quality ... as the Japanese Navy could not compete with the United States Navy in regard to the number of battleships, it was decided to construct the two largest and most effective battleships in the world, the *Yamato* and *Musashi*.[19]

The Plan called for two new battleships, two aircraft carriers, 15 destroyers, 50 submarines, one large and five small minelayers, 27 other vessels, 262 land-based and 294 ship-borne aircraft.[20] The only ships where quantity remained a consideration were the vulnerable aircraft carriers, and with these the Japanese decided that they had to match US numbers at least, accepting that these vessels would be prime targets and likely to be lost. As there was a shortage of naval building slips, the only way of achieving these numbers was by converting suitable merchant ships, and suitable hulls were selected for conversion. The total budget for the Replenishment Plan was some 806,549 yen in 1936, with an additional 806,549 yen between 1937 and 1941.

The involvement of the Imperial Japanese Navy in the China Incident was another factor in the establishment of subsequent plans which were were passed by the Japanese Diet:

> Believing that the Incident would continue to expand for a considerable time, and that friction with the Three Powers would necessarily increase during that period, the Navy determined to accelerate war preparations immediately to cope with the situation.[21]

When, in 1934, the United States started their own fleet expansion, the Japanese response was the Fourth Replenishment Plan, which aimed for the accelerated completion of surface units, the requisitioning of merchant ships for conversion to aircraft carriers and the doubling of the number of naval aircraft, and also 95 additional vessels. This would make Japan the strongest naval power in the Pacific by 1945, and more powerful than either the United States Navy or the Royal Navy, even with its 'New Standard Navy', could hope to match, according to Captain Tufnell, the Tokyo Naval Attaché. As Backhouse commented in 1939, short of an economic setback for Japan, Great Britain would be in a serious situation and would lose naval supremacy to Japan by 1945.[22]

Understandably, given the Japanese obsession with secrecy, British estimates of Japan's intentions were wide of the mark. Tufnell believed that Japan could complete between four and eight capital ships by 1945, which, when added to the existing fleet of ten ships, would give little room for complacency to a Royal Navy which had a battle fleet of 15 ships, with as many as six laid up, in rotation for refit and modernization, and the five *King George V* class building. In a worst-case scenario the Royal Navy would thus have only seven battleships available to oppose the Japanese fleet and, in fact, when the Japanese proposed their fifth and sixth Replenishment Plans, which aimed, on paper at least, to counter US plans for a two-ocean navy, British expansion plans were being definitely overshadowed.

The British were concerned about the growth of Japanese naval power, about their own limitations regarding the projection of naval power into the region, and also about the Japanese obsessive secrecy concerning their naval construction programmes, to the extent that in the 1933 edition of *Brassey's Naval and Shipping Annual*, Captain E.H. Althan, RN, wrote:

> Details of Japanese naval construction released for publication are meagre, but there appears to be a strong tendency to build ships more powerfully armed and better protected, and to obtain these advantages at the expense of radius of action. This seems to indicate that, in the event of war, Japan would not seek to join action in enemy waters but would prefer to engage a hostile fleet nearer her own ports, where she would be at a considerable advantage strategically, and possibly even tactically – a very logical policy for a nation without overseas naval bases.[23]

If this was to be the case, and it was the most likely scenario, then it would have a great influence on how the British would have to respond and helps to explain why War Memorandum (Eastern) may have shifted from the earlier versions, which emphasized a fleet-to-fleet encounter, to one which both acknowledged the British difficulties in achieving this and which emphasized a war against

Japanese trade, aiming to place the Japanese on the defensive and draw the battle fleet out into a decisive action. The increase in Japan's naval strength can be seen in Table 9.

Table 9
Japanese Naval Strength

End 1924	End 1930
6 Battleships	6 Battleships
4 Battle cruisers	4 Battle cruisers
2 Aircraft carriers	4 Aircraft carriers
18 Heavy cruisers	18 Heavy cruisers
13 Light cruisers	21 Light cruisers
30 1st Class destroyers	56 1st Class destroyers
51 2nd Class destroyers	48 2nd Class destroyers
30 3rd Class destroyers	—
50 Submarines	67 Submarines
7 1st Class coast defence ships	7 1st class coast defence ships
4 2nd Class coast defence ships	2 2nd Class coast defence ships
plus lesser warships and auxiliaries	plus lesser warships and auxiliaries
Total 213 ships	270 ships

Source: 'Outline of Naval Armaments and Preparations for War, Part I' and Chesneau (ed.), *Conway's All the Worlds' Fighting Ships, 1922–1946*.

By 1934, Japan had built up a large fleet, despite its treaty obligations, and in 1936 *Brassey's* commented on the extent of Japanese strength; with the exception of *Hiei*, which was still demilitarized, nine capital ships were judged to be as powerful as the 15 Royal Navy ships, owing to their extensive refits and modernization. Japan also had four aircraft carriers afloat and two more, *Soryu* and *Hiryu*, laid down, in addition to a very powerful cruiser force and a heavily armed destroyer fleet.[24]

The problem for the Japanese after Washington was how to contend with equal or superior naval forces and the annual operational plan gave increasing attention to reducing an enemy's strength through continued attacks by light forces. As with contemporary British tactical thinking,

> it was left to the overall Japanese commander to decide where and when to engage the enemy in the final showdown, a decision based on the results of the attrition operations, weather conditions and a number of other variables. But in the 1920s, at least, it was assumed that the decisive fleet battle would take place during the day, probably in the first hours of daylight after a long night of attacks by light forces.[25]

Later, Japanese expertise in night actions, to make these attacks more decisive, was developed and refined to the point where the Japanese became renowned for their night-fighting proficiency.

Like the British and the Americans, the Japanese also held on to a continued belief in the naval gun battle as the decisive fleet encounter, involving all types of warship, but centred on the battleship. For years this was the subject of meticulous study at the Japanese Naval Staff College and the focus of periodic revisions of the Battle Instructions, and other ways of defeating an enemy, for example, through attacks on commerce, were ignored or deemed to be of secondary importance.[26] The Imperial Japanese Navy's development of submarines illustrates this well. They were large, some were fitted with a catapult and aircraft and were to be employed against the enemy's naval forces as a way of equalizing the numbers and to make up for Japan's smaller number of battleships, whilst all of Japan's naval vessels were to be built as heavily armed as possible as a way of similarly making up for lack of numbers.

Because of the limitations on building battleships in the 1920s and 1930s, the interception plans were based around cruisers, destroyers, submarines and aircraft. In 1924 the Japanese had started to organize a squadron of torpedo cruisers for night raids and the light cruisers *Kitakami* and *Oi* were rebuilt to carry ten quadruple torpedo mounts, giving them the heaviest torpedo armament of any ship in the world. In 1929 the command and responsibility of all night raids was assigned to a heavy cruiser squadron under the commander of the Second Fleet, the fleet that would conduct these interception moves.[27]

Japanese battleships were also modernized, with improved boilers, better armoured protection and increased gun elevation, up to 43°, and new cruisers and destroyers were authorized and built.[28]

Japanese cruiser development reflected the quest for a versatile vessel, intended in the Imperial Japanese Navy partly to make up for the lack of battleships and partly to maintain the operational principles of the 8:8 Plan. The Imperial Japanese Navy developed fast, heavily armed cruisers, able to perform a range of tasks, starting with the *Furutaka* class, laid down in 1922 and armed with six 8-inch guns and 12 torpedo tubes. They were followed by the four bigger ships of the *Myoko* class, armed with a main battery of ten 8-inch guns, able to steam at a maximum of 35 knots and with a range of 10,000 miles. With this combination of high speed and massive firepower they were able to enter or avoid combat as circumstances dictated, and although initially unstable, they were to prove formidable opponents, and would have probably far outclassed any single Royal Navy cruiser.

They were joined by the next class, the *Takao* class, and their role was identical: they could use their speed to keep in contact with enemy forces, withdrawing out of range when threatened by the enemy screen. At nightfall they would join with the destroyers and the specially converted torpedo cruisers *Kitakami* and *Oi* and use their firepower to smash through the enemy's defensive cruiser and destroyer screen, adding their own torpedoes to the salvoes launched against the enemy's confused battle line, before withdrawing to rejoin their own battle fleet and form a fast division, ready to act against the enemy van, or to attack and sink any damaged, enemy stragglers.[29]

As the Navy General Staff had decided that one good way of enhancing the Japanese capability of winning a decisive surface battle was to build as many ships as possible capable of launching torpedoes, allowing the Japanese battle fleet to contend with a superior enemy fleet, it made sense to build destroyers with the heaviest torpedo and gun armament possible. Although the annihilation of the enemy fleet in the decisive, surface naval battle was still to be the main aim of the Japanese, reducing the size of their opponent's fleet by a series of night torpedo attacks was just as important.

The *Fubuki*-class destroyers of 1923 were their response to the need for a heavily armed destroyer. Each ship carried six 5-inch guns, nine torpedo tubes and three reloads for each tube, something no other navy had, giving them the heaviest torpedo salvo of any destroyer and indicating that this was a specialized vessel, not the 'maid of all work' that destroyers in the Royal Navy tended to become.

The Japanese overall operational plan also called on their submarine flotillas to shadow the enemy fleet, and the large *I* boats, of 14,000 tons displacement, capable of making 20 knots on the surface and 10 knots submerged, and carrying an armament of nine torpedo tubes, were designed with this in mind. The Japanese decision to deploy submarines with the battle fleet largely ignored the German experience in the First World War and that of the British and the *K* class; the Pacific was a different theatre, with unique conditions, and Japanese priorities would be different, more concerned with attacking the United States Navy crossing the Pacific, or the Royal Navy moving northwards from Singapore, than acting against their merchant trade, which was not expected to be numerous anyway, once a war had been declared. By 1924 the Japanese had two divisions of submarines in commission, and were developing large, long-range or 'fleet' and 'cruising' submarines, both designed to operate in flotillas against an enemy battle fleet.

It was therefore the battleship that was to play a leading role in Japan's decisive daylight naval battle, although the cruisers, destroyers and submarines, with naval air power, would achieve a measure of

14. With an armament of eight 16-inch guns, *Mutsu* (seen here) and her sister ship, *Nagato*, were the largest battleships the Japanese possessed until the launching of the first of the *Yamato* class in 1941. The British expected to be outranged and outgunned, but not outclassed, if they met *Mutsu* in battle.

15. The battleship *Fuso*, with 12 14-inch guns and a speed of 24 knots was the equal of the *Queen Elizabeth* class and was reckoned, by both sides, to have been able to give a good account of herself in any fleet engagement.

16. The fast battleship *Kongo*, armed with eight 14-inch guns. She and her sisters would have been equal to the *R* class in battle, and had a clear superiority in speed over the *Queen Elizabeth* and *Nelson* classes.

17. The aircraft carrier *Kaga* was originally laid down as the name ship of a new class of battleship which was due to be scrapped under the terms of the Washington Treaty. She was reinstated into the Fleet programme and converted to an aircraft carrier in 1924. Like the Royal Navy's aircraft carriers, she originally had three flying off decks forward. Her long vents for funnel gases are clearly visible.

18. With her aircraft catapult clearly visible, forward of the bridge, the light cruiser *Yura* was intended to scout for the Japanese battle fleet and lead flotillas of destroyers into attacks on the enemy battleships.

19. Improved versions of Japan's first Treaty cruisers, the *Furutaka* class, *Aoba* and her sister ship *Kinugasa*, mounted six 8-inch guns. The launching of the *Furutaka* and *Aoba* classes prompted replies from the Royal Navy and United States Navy and led to cruisers being the centrepiece of naval negotiations during the inter-war years.

20. One of the first four Japanese cruisers to exceed 10,000 tons, *Ashigara* mounted ten 8-inch guns, and originally had eight 4.7-inch guns and 12 fixed torpedo mounts. They would have been formidable opponents in a fleet action.

21. Typical of Japan's inter-war destroyers was *Yusuki*, armed with four 4.7-inch guns and six torpedo tubes. The *Mutsuki* class, of which *Yusuki* was a member, were the first ships to carry Japan's deadly, 24-inch 'Long Lance' torpedo.

22. When first launched, *Shikinami* and her sister ships of the *Fubuki* class outclassed any other navy's destroyers. They originally carried six 5-inch guns in dual-purpose mounts and nine 24-inch torpedoes, with a range of 43,000 yards, along with a further nine spare torpedoes, which could be reloaded and fired within 15 minutes.

superiority through a long period of attrition by the light forces, as the enemy fleet moved nearer to Japan.

It is probable that by the late 1930s the Imperial Japanese Navy had adapted the operational principles originally enshrined in the 8:8 Plan and was, like the Royal Navy, training its battle fleet (*Fuso, Yamashiro, Ise, Hyuga, Nagato* and *Mutsu*) to fight in two, three-ship divisions, each one supported by two divisions of cruisers and two destroyer flotillas. This force supported the vanguard force of the four fast battleships (*Kongo, Kirishima, Haruna* and the re-armed *Hiei*), three to four divisions of cruisers and two destroyer flotillas. Long-distance torpedo attacks would be launched by the advanced division at about 35,000-yard range, and the ships would then fall back on the battle fleet for support. Once the battle fleet was at about 35,000 yards range from the enemy, they would commence long-range firing. (The Royal Navy, by comparison, was expecting to open fire at a range of about 20,000 yards.) As with the British tactical plans, it was anticipated that torpedo hits and gunfire hits would start to throw the enemy line into confusion, making it harder for them to hit back, as the Japanese closed in to around 21,000 yards and commenced an intensive bombardment. At this point, more destroyers would close in, to a range of 5,000–6,000 yards, and launch more spreads of torpedoes. In theory, each destroyer division was reckoned to be able to at least damage one enemy battleship, but in the *mêlée* which would ensue, with destroyers and cruisers fighting each other between the two battle fleets, streaming smoke, throwing up large bow waves, firing at each other, and wrecks burning and sinking, as well as the heavy cruisers wheeling around the van and rear, this was an over-optimistic assessment. Japanese studies of the Battle of Jutland had revealed to them, as to the British, that to be effective in a night action, a battle fleet had to be properly led, skilfully deployed and highly trained, so that throughout the 1920s the Japanese continually practised and refined their night-fighting tactics with just this in mind.[30]

Missing from the above is any mention of the contribution that the Japanese expected their aircraft carriers and embarked aircraft to make, and it is in this arm that the Imperial Japanese Navy's operational doctrine fundamentally differed from the Royal Navy's.

Although the Japanese later achieved great success with their naval air arm, principally the attack on Pearl Harbor and the sinking of *Prince of Wales* and *Repulse*, the development of an air arm for the Imperial Japanese Navy was initially slower than in the Royal Navy. It was not until 1916, with the formation of the Yokosuka Air Group, that the Japanese acknowledged the potential of aircraft at sea and developed a naval aviation arm. The Sasebo Air Group followed in

1918, and in the same year a plan was adopted to form more units, bringing the total to eight, and, in 1920, further plans were made to extend this with another nine air groups. The launching of the *Hosho* and the assistance of the British Sempthill Mission showed that the Japanese were as committed as the British to integrating air power into naval warfare, and although the planned increase to 17 air units was delayed until 1930, air power was recognized as one of the ways that Japan could negate the Washington Treaty and improve its naval forces' chances of inflicting serious damage on any enemy battle fleet.[31]

Initially, the Imperial Japanese Navy and the Royal Navy saw very similar roles for their naval aircraft; as in the Royal Navy, the larger Japanese aircraft carriers would be operating at some distance from the battle fleet, with the battle cruisers and cruisers, trying to locate the enemy aircraft carriers, and destroy them to achieve air superiority for the Japanese, whilst the smaller aircraft carriers operated with the battle fleet, providing aircraft for spotting and reconnaissance, as well as fighter defence and a limited strike potential. Later, as Japan's navy gained more expertise in naval aviation, this role would evolve.[32]

By the mid-1920s, the Japanese had begun to develop both land-based, and sea-based, naval air arms and, as in the Royal Navy, by the end of the decade, aircraft were being used for scouting, reconnaissance, fleet air defence and spotting. As in both the Royal Navy and United States Navy, early Japanese aircraft carrier designs reflected the uncertainties about the use of aircraft carriers with the battle fleet. Offensive naval aviation was not yet a fully developed capability and both the *Akagi* and *Kaga*, in common with their US counterparts, *Lexington* and *Saratoga*, originally carried 8-inch guns to defend themselves against enemy cruisers. Whilst this appeared logical as long as aircraft carriers were expected to operate with the battle fleet and therefore within range of the enemy's guns, exercises showed that engaging in a gun duel, whilst loaded with highly inflammable aviation fuel, was simply suicidal. In fact, it was the recognition of the vulnerability of aircraft carriers that decided the Japanese to look to mercantile conversions rather than build aircraft carriers from the keel up; they expected to lose aircraft carriers and in this case wanted quantity, not quality. And with no spare slipways, Japanese naval shipbuilders just could not build aircraft carriers in sufficient numbers from the keel up, anyway.[33]

A distinctive Japanese aircraft carrier doctrine only began to be fully developed in 1928 when the First Carrier Division, *Akagi* and *Hosho*, was formed and a serious study of the role of aircraft carriers with the fleet was begun. And as in the Royal Navy and United States Navy, the short range of carrier-borne aircraft still limited them, initially, to supporting the battleship in the decisive action, spotting and maintaining air superiority to deny the enemy spotters opportunities to report on the Japanese ships. Air strikes would be launched against the enemy carrier force, to deny them this tactical advantage. However, as aircraft improved, the increased range and effectiveness of aircraft led to Japanese gunnery and air officers realizing that the pre-emptive air strike against an enemy could damage the enemy battle line before the two battle fleets sighted each other, whilst destroying the enemy aircraft carrier force and winning air superiority over an enemy fleet would greatly disadvantage the enemy battle fleet. To do this the Japanese would need to locate an enemy fleet, over the horizon, before they were sighted themselves, which led to the integration of floatplane reconnaissance from battleships, cruisers, seaplane tenders and even submarines. And if aircraft carriers were going to conduct strikes early on, they had to be freed from protecting the battle fleet. But dispersing the carriers, in its turn, made them more vulnerable to enemy attack, unless accompanied by heavy ships such as cruisers and battleships.

In an unpublished monograph, Minoru Genda, who later achieved fame as one of the leaders of the raid on Pearl Harbor, claimed that Japanese tactical doctrine for naval aviation developed in five phases. Phase 1, from 1923 to 1928, was basically when the Japanese, with the help of the British, studied and trained in the use of the aircraft carrier, using first *Hosho*, and then both *Hosho* and *Akagi*. Numbers of ships and aircraft remained small, however, and the tactical use of aircraft reflected the British influence: aircraft carriers being a part of the battle fleet, the aircraft being used mainly for reconnaissance, torpedo and horizontal bombing attacks on enemy ships, and defence of the Japanese line of battle. As with the Royal Navy, aircraft were generally expected to find and attack the enemy aircraft carriers first, in order to deny their opponents the use of their own aircraft, but then they would perform essentially subordinate roles to support the battle fleet.[34]

Despite the shortage of money caused by the world-wide depression during the 1930s, the Imperial Japanese Navy did conduct exercises. In March 1929 the Combined Fleet paid a courtesy visit to northern China, exercising *en route*, and further joint manoeuvres were held in April, May and June, during visits to Tsingtao and Chemulpo. Mention of the use of aircraft is prominent in the British naval attaché's 1929 Naval Notes, those for May noting the use of catapult-launched aircraft aboard the *Furutaka*-class cruisers of the

5th Cruiser Squadron, commenting on the growing importance of naval air power throughout March to June, and a report from *Nichi Nichi*, of 22 June, quoting Rear Admiral Takahashi about future aircraft carrier plans:

> The aircraft carrier *Kaga* will be commissioned in December and relieve the *Akagi*. It is considered very inconvenient that she will form a squadron with the *Hosho* which is much inferior in speed, but it is very convenient for the investigation of comparative efficiency of 27,000 tons and 12,000 tons aircraft carriers, which is now widely discussed. The American navy is in favour of the small aircraft carrier and has laid down 13,000 ton ships, but I am of the opinion that this question could not be so easily settled. The present day science has practically exhausted all its resources in the construction of aircraft carriers but there is still plenty of room for study of the aeroplane.[35]

This was at a time when navies were looking at ways of getting the biggest and best fleets from the treaty limitations; smaller battleships, small aircraft carriers, hybrid battleships and cruisers were all being seriously investigated by all three major navies at the time. The size of aircraft carriers was subject to a great deal of debate, especially in the United States Navy, which initially did not like their large carriers, *Lexington* and *Saratoga*, feeling that they were too vulnerable, concentrated too much naval air power in one ship and tied up too much of their tonnage. In the Royal Navy, with its fleet of six aircraft carriers, however, *Argus*, *Eagle* and *Hermes* were regarded as too small and were to be replaced by larger ships as soon as possible. Takahashi seemed to be inferring that the Japanese were still considering whether larger or smaller aircraft-carrying ships, possibly including hybrid designs but all capable of carrying modern aircraft, would suit Japanese strategical and tactical doctrine. A further report in the 8 August edition of *Hochi Shimban* reported on the laying down of the new aircraft carrier *Ryujo*, pointing out that, although smaller than the *Kaga* and *Akagi*, it had none of their defects and would be, in the newspaper's view, the 'finest aircraft carrier in the world'.

Japanese fleet manoeuvres carried out in September 1929 concentrated on the use of aircraft at sea. The 1st Aircraft Carrier Squadron (*Akagi*, *Hosho*, seaplane tender *Notoro* and the 4th Destroyer Flotilla) took on the role of invaders, attacking the Bonin Islands before moving on to assault Tokyo Bay, where it was expected that a decisive naval battle would take place. The defenders were the land-based aircraft from the *Yokosuka*, *Kasumigara*, *Sasebo* and *Omura Flying Corps*, assisted by the 7th Destroyer Flotilla.

According to one newspaper report, from the Tokyo *Nichi Nichi*, compiled by the British naval attaché into a report for the Admiralty, the attacks began on 23 September when 12 planes from the aircraft carriers appeared over Oppama, whilst in return the aircraft carriers were located and bombed, all three ships being judged to be disabled by the judges. And in a similar move on 24 September, all of the raiding aircraft were judged to be destroyed after a fierce air battle over Oppama. Both of these exercises seemed to show the superiority of the Japanese defensive arrangements over an enemy's attacks and the need to develop attacking procedures further. Doing this would mean a greater investment in aircraft carriers and their air groups; in other words, greater expenditure on the navy.

More large-scale exercises were carried out in 1930. In February the naval attaché's reports mentioned successful bombing exercises from the aircraft carriers, whilst in March, a surface bombardment of Tokyo was judged to have been repulsed by shore batteries. Significantly, also, in February there is mention of successful night deck landings being carried out on *Akagi*.

October 1930 saw further exercises. Ships of the Combined Fleet (*Mutsu*, *Haruna*, *Yamashiro*, *Ise*, the aircraft carriers *Kaga* and *Hosho*, with supporting cruisers and destroyers), 70 warships, supported by the 1st and 2nd Air Fleets, were designated as the Blue Fleet, tasked with defending Tokyo Bay, whilst the Third Fleet (battleships *Nagato* and *Haruna*, aircraft carrier *Akagi*, and 60 older ships) was to be the Red Fleet, and was to assemble at Ise Bay, seize Honshu, and then try to draw the Blue Fleet away from its assembly point at Kure and into Honshu waters, where it would be attacked by Red destroyers and submarines.

Another series of exercises began on 21 October, involving all ships from the Combined Fleet and Reserve Fleet. The Combined Fleet, with 15 transports, would now be the invader, attempting to land 20,000 troops of the 35th Brigade on the northern coasts of Kyushu, whilst the Third Fleet, made up of ships normally in reserve or designated for coast defence, had the task of trying to destroy the invasion fleet.[36] The *Japan Times*, Osaka *Mainichi*, Sasebo *Gunko Shimban* and Tokyo *Nichi Nichi* all carried reports of this exercise, which consisted of fierce destroyer engagements, some at night, air reconnaissance, torpedo bombing and level bombing, including the simulated use of incendiary and poison gas bombs against shore installations. Fighting was reported to be at close range, and there were reports of the Red submarine force and aircraft cooperating against the Blue forces. The decisive battle was fought off Shosaki on the morning of the 21 October, and the defending Blue fleet was judged to have lost. Red's superior use of destroyer attacks, aircraft and submarines were all lessons that the Japanese navy was going to study further, and the Kure *Nichi Nichi* quoted Admiral Taniguchi, of the Naval General Staff, on his reactions to what was, after all, a simulated defeat of Japan's defences:

> All the umpires and commanders are studying to find out what is responsible for the defeat. They are more than amazed to discover the efficiency of the Air Arm in sea fighting and how hard it is to face an enemy with a superior air force.
>
> The efficiency of submarines as defence arms and their effect on tactics, and the efficiency of the 10,000 ton cruisers were also surprising.
>
> The defending side used destroyers to the fullest extent to attack the enemy, but it was almost impossible to face the attacking force which had a superior air force, 8-inch guns etc.[37]

In short, the Imperial Japanese Navy was conducting its own, vigorous investigations in how best to integrate the use of aircraft into its tactical doctrines, to enable it to reduce the enemy's numbers on the approaches to Japan and once decisive battle was joined, to enable the battle fleet to fight and win a decisive battle. Exercises often only show the results that the planners wish them to show and a 'defeat' for the defenders of Japan would undoubtedly strengthen the case for greater amounts of money for naval expansion, especially of the naval air arm, which was a relatively inexpensive way for Japan to make up for its numerical lack of capital ships when compared with the United States Navy and Royal Navy.

So, when the London Naval Conference failed to give Japan their hoped-for 70 per cent margin of superiority in cruisers, a more developed set of operational tactics for naval aircraft emerged. Carrier-borne naval air increasingly began to play a major part in Japanese naval thinking after the *Kaga* was commissioned in 1930, giving Japan two large and one small aircraft carriers. Land-based naval aircraft, including bombers and torpedo bombers, however, were frequently included in tactical planning and exercises.

Japanese aircraft carriers were now expected to adopt a more aggressive role, and to make surprise attacks on their enemy counterparts. Aircraft carriers began to exercise, not just with the cruisers of the advanced screen, but also as independent squadrons, not as a part of the battle fleet, whilst dive-bombing tactics, with their key elements of speed and surprise, began to be developed. Between 1935 and 1940 this independent role, aimed at surprising and sinking enemy aircraft carriers, became the most important of the Japanese aircraft carriers' roles and more tactics began to be developed for operating the aircraft carriers as separate squadrons.[38] Such tactics also suited the Japanese aggressive mentality, as well as emphasizing the aircraft carrier's greater potential as an attacking vessel. Unseen and unlocated, it was a formidable weapon. Detected, it was a large, volatile target for enemy bombs and torpedoes.

At about the same time the Japanese, using their combat experience in China, decided that bombers had to be escorted by high-performance fighters, and that the attacking formations had to be as large as possible, in order to ensure a successful, overwhelming attack on an enemy. Pre-emptive strikes also called for a massed air strike, which allowed for a larger combat air patrol to supplement the fleet's anti-aircraft protection. However, the launching of a multiple strike presented other problems to the Japanese, notably, how to gather and concentrate the large formations, and how to launch them at an effective strike distance without having their own carriers detected. The solution was to operate the aircraft carriers as one squadron, working in pairs or fours, not only to ease concentration of aircraft, but also to aid the aircraft in finding their aircraft carriers after a raid.

Such formations, arriving over an enemy target at the same time, and launching a series of coordinated, but very different, attacks, would gain air supremacy over an enemy fleet, overwhelming any defensive fighters, strafing open decks and, with a combination of high-level bombing, dive-bombing and torpedo bombing, reduce the effectiveness of the anti-aircraft barrage. The concentration of the aircraft carriers that this would inevitably entail would also allow for a stronger air and anti-aircraft defence of the aircraft carriers and supporting ships.

The solution to launching the necessary large air strikes was to concentrate the aircraft carriers into one squadron, and as the potency of the aircraft was fully recognized, more cruisers and battleships were attached to the aircraft carrier squadron, with the express aim of seeking out damaged enemy ships and sinking them by gunfire once they had been disabled by the aircraft attacks. This was the type of fleet which attacked Pearl Harbor, which ranged into the Indian Ocean in 1942, and which fought at the Battles of the Coral Sea and at Midway, but which again differed fundamentally from the British tactical use of aircraft carriers, despite the exercises carried out in 1931 in the Mediterranean by Rear Admiral Henderson.

From an examination of some of the available evidence it is therefore possible to suggest a likely Japanese strategy that would have been used against the British, and how a battle may have unfolded. No concrete war plans for a war with Britain were actually written, only general principles developed, which reflected Japanese strategical ideas:

> The main points added up to these: in the first phase, the Navy would smash British naval forces in the East by a swift strike, at the same time co-operating with the army to capture British Borneo and key areas on the east coast of Malaya, also Singapore and Hong Kong. In the second phase, they would annihilate the British main fleet on its arrival in eastern waters. The 1940 plan stated: 'In case the main force of the enemy tries to hold out based on India or Australia we should make efforts to lure the enemy out by diminishing his strength and destroying his line of communications while securing control of the sea.'[39]

The strategic imperative remained unchanged throughout the period between the two world wars: to protect Japanese trade and to bring the enemy fleet or fleets to battle. In order to achieve this, an outer defensive ring was to be established, enemy forces in the region neutralized and their trade attacked. This was actually easier during 1941 than it would have been if a war had been fought, say, in 1925, when the British initially developed war plans. By 1941 the Japanese controlled the coast of China, had air bases in Thailand and French Indo-China, and had developed bases on their Pacific possessions. These, along with the attacks on Malaya, the Philippines, the Netherlands East Indies and various Pacific islands gave Japan, first, air bases to support its other operations, and then, by 1942, its early defensive ring and a secure supply of raw materials.

In short, by the outbreak of war in 1941, the Japanese were supremely confident in their ability to fight and annihilate any forces the Royal Navy could position in the Far East. Through the sinking of the battleship *Prince of Wales* and the battle cruiser *Repulse* and the defeat of the combined Australian, British, Dutch and US cruiser force at the Battle of the Java Sea, they did just this. US naval power had already been neutralized by the successful attack on Pearl Harbor, and the raid into the Indian Ocean in 1942, which effectively neutralized Admiral Somerville's force of older battleships, forced the Eastern Fleet onto the defensive. And although they failed to engage Somerville in a decisive battle, Japanese naval superiority was complete, at least for a period of time.

Tactically, any British battle fleet fighting the Japanese during the 1920s and 1930s would have expected to have incurred losses from mines, torpedoes and destroyer attacks from the time it approached Singapore for the first time, until it moved nearer to the Japanese trade routes with the Chinese mainland, within range of Japanese land-based aircraft. By the 1930s, with a changing European situation, when it was becoming increasingly unlikely that a British battle fleet could be sent eastwards at all, Britain was looking more and more towards the United States to provide the ships, offering them the use of Singapore for operations in defence of the Philippines and the Malaya Barrier, against the Japanese.

NOTES

1. M. Peattie, 'Japanese Naval Construction, 1919–41', in O'Brien (ed.), *Technology and Naval Combat*, p. 93.
2. A. Temple Patterson (ed.), *The Jellicoe Papers, Vol. II* (Navy Records Society, London 1968), pp. 284–397; Tracy (ed.), *The Collective Naval Defence of the Empire, 1900–1940*, pp. 241–4.
3. G. Freidman and M. Lebard, *The Coming War with Japan* (St Martins Press, New York, 1991), p. 519.
4. NHB, London, 'Naval Staff History: War with Japan, Vol. I'.
5. Marder, *Old Friends, New Enemies*, pp. 23–4.
6. D. Evans and M. Peattie, *Kaigun: Strategy, Tactics and Technology in the Imperial Japanese Navy, 1887–1941* (Naval Institute Press, Annapolis, MD, 1997), p. 143.
7. NHB, 'Naval Notes 1921: notes prepared by the Naval Attaché, Tokyo', February–November 1921.
8. Ibid.
9. Evans and Peattie, *Kaigun*, pp. 197–203.
10. Ibid., ch. 1, pp. 199–237; R. Chesneau (ed.), *Conway's All The World's Fighting Ships, 1922–1946* (Conway Maritime Press, London, 1980), pp. 167–217.
11. With no base in the Philippines able to support battleships, the United States Asiatic Fleet consisted of cruisers, destroyers, submarines and gunboats. With a base but no fleet, the Royal Navy's China Squadron typically comprised one small aircraft carrier – *Eagle* or *Hermes* – cruisers, destroyers, submarines and gunboats.
12. Peattie, 'Japanese Naval Construction', in O'Brien (ed.), *Technology and Naval Combat*, p. 97.
13. NHB, 'Naval Notes 1929: notes prepared by the Naval Attaché, Tokyo', January 1929.
14. Ibid.
15. Ibid.
16. M. Peattie, *Sunburst: The Rise of Japanese Naval Air Power, 1909–1941* (Naval Institute Press, Annapolis, MD, 2001), pp. 27–37.
17. NHB, 'General Headquarters, Far East Command, Military History Section; Special staff, Japanese Research Division; Outline of Naval Armament and Preparations for War', Monograph 145 (1945), pp. 6–7.
18. Ibid.
19. Ibid., Monograph 149, p. 2.
20. Ibid., Appendix 1.
21. Ibid., p. 8.
22. Marder, *Old Friends, New Enemies*, pp. 16–17.
23. C. Robinson and H.M. Ross (eds), *Brassey's Naval and Shipping Annual, 1931* (W. Clowes and Sons, London, 1931), p. 36.

24. C. Robinson and H.M. Ross (eds), *Brassey's Naval and Shipping Annual, 1936* (W. Clowes and Sons, London, 1936), pp. 33–5.
25. Evans and Peattie, *Kaigun*, p. 204.
26. Ibid., p. 212.
27. Y. Hirama, 'Interception–Attrition Strategy from the Washington Treaty to the Battle of the Philippine Sea', in *Les Marines de Guerre du dreadnought au nucleaire* (Actes du colloque international, Paris, ex-Ecole Polytechnique, les 23, 24 et 25 novembre, 1988, Service historique de la Marine), p. 397; Chesneau, *Conway's, 1922–1945*, p. 174.
28. Ibid., pp. 171–4.
29. Ibid.
30. Evans and Peattie, *Kaigun*, pp. 282–8.
31. C. Robinson and H.M. Ross (eds), *Brassey's Naval and Shipping Annual, 1932* (W. Clowes and Sons, London, 1932), p. 40.
32. M. Genda, 'Evolution of Aircraft Carrier Tactics in the Imperial Japanese Navy' (unpublished monograph).
33. NHB, Monograph 160, Part III.
34. Genda, 'Evolution'.
35. NHB, 'Naval Notes 1929; notes prepared by the Naval Attaché, Tokyo', June 1929.
36. Ibid., September 1929, October 1929.
37. Ibid., October 1930.
38. Genda, 'Evolution'; Peattie, *Sunburst*, pp. 147–53.
39. Marder, *Old Friends, New Enemies*, pp. 325–6.

8

Main Fleet to Singapore: The Sinking of HMS *Prince of Wales* and HMS *Repulse* and the End of War Memorandum (Eastern)

The surprise Japanese attack on Pearl Harbor on 7 December 1941 removed a key element of Britain's revised Far Eastern war plans, the United States Pacific Fleet. After two years of war it had become obvious to the British that only by either persuading the United States to station its Pacific battle fleet at Singapore, the only fully developed naval base west of Hawaii, or convincing President Roosevelt's administration to send more vessels into the Atlantic Ocean to escort convoys and provide cover against the remaining German battleships could the Royal Navy concentrate the necessary ships to send to the Far East. At a stroke, the Japanese carrier-borne aircraft had removed this option, and the small British squadron sent to Singapore was now dangerously exposed, isolated and vulnerable, in exactly the position that the pre-war planners had predicted for a 'flying squadron' without the support of the main fleet.

The Royal Navy had been through the difficult times of the mid-1930s when a combination of the re-emergence of Germany as a European military power, the Abyssinian Crisis and the Spanish Civil War, coupled with the constitutional crisis brought about by the abdication of King Edward VIII and growing pacifist feeling throughout the country, had all combined to create a general climate of uncertainty and a lack of confidence in the ability to wage war. However, the work of Chatfield, when First Sea Lord, in pressing for a programme of naval rearmament, which included signing the Anglo-German Naval Treaty as a way of buying time whilst rearmament gathered pace, seemingly had halted, even for a short time, the threat from growing German naval expansion, whilst the political developments in the Mediterranean at the time of the Abysinian Crisis at least had revealed some of the defensive problems there, even if they had been barely rectified by the outbreak of war in 1939.

In the Far East more work was being undertaken in Malaya to develop the defences of Singapore, including new airfields at Tengal and Sembawak and a further five elsewhere in Malaya. However, naval policy remained fundamentally the same – the defence of home waters and the Mediterranean; the vital lines of communication to the oilfields of the Persian Gulf, and to the Dominions, assumed a greater priority than the defence of Singapore and Britain's Far Eastern possessions.

This was despite the deterioration, since 1934, of the situation in China, where the Japanese had directly challenged the British, Dutch and US presence in the region, taking over responsibility for keeping the peace, a move designed to establish a Japanese autarky and safeguard the Japanese economy. The following year saw the crisis in Shanghai, which started over the alleged kidnapping, by the Chinese, of a Japanese deserter, and quickly escalated to the point where the Japanese had 20 cruisers and destroyers supported by heavy units off the mouth of the Yangtze.

All the British could station in the region was the China Squadron, on paper an impressive fleet, numbering over 100 vessels and including the aircraft carrier *Eagle*, three heavy and two light cruisers, a destroyer flotilla, five escort sloops, 17 submarines, 18 gunboats and various other smaller craft and auxiliaries, enough to represent Britain's interests, but nowhere near strong enough to oppose the Japanese Navy. In the circumstances, the British government had no wish to antagonize the Japanese government, and advised the masters of British ships to submit to searches by the Japanese looking for contraband.[1] This was, after all, not a war, merely an 'incident'.

Similarly, when British sailors were beaten up by Japanese police at Keelung in Formosa, all that the British felt able to do in retaliation was to end warships' port visits to Japanese territories, denying themselves a valuable source of intelligence on Japan's ships. When the destroyer *Decoy* was nearly rammed by a Japanese cruiser escorting merchant ships, the Admiralty warned warships not to manoeuvre near Japanese warships in a way that might be taken as warlike, whilst the Japanese Admiral Hasegawa ominously suggested no salutes be fired, to avoid misunderstandings.[2] And when a sailor from the *Decoy* got drunk at Tsingtao and defaced the Japanese flag, the Captain had to secure his release by going ashore in full ceremonial regalia and apologizing.

Other incidents included the Japanese firing on and bombing British and US gunboats. Once Canton had been captured, the Japanese closed the trade of the Pearl River to foreign shipping, whilst the closure of other ports such as Tsingtao created a virtual

Japanese monopoly in China. Matters came to a head at Tientsin in 1939, when the Japanese seized a consignment of Chinese silver being shipped to Shanghai and demanded the British and French authorities hand over the rest, still in Tientsin. In April they blockaded the Concessions after four alleged Chinese terrorists sought sanctuary there. By July there was near famine and only when the four detainees were handed over to a puppet Chinese court did the Japanese lift the blockade.[3] Britain was powerless, something that the Japanese could not have failed to have noticed.

In Europe, the invasion of Czechoslovakia in 1939 made war inevitable. No war plans could ignore the Japanese, however, and in an effort to gain up-to-date information about the Japanese, the cruiser *Birmingham* made an unannounced voyage to Amoy and photographed two of the four lines of Japanese ships there before equipment was hastily covered up. Nevertheless, the China Squadron was being stretched to the limit representing Britain's interests, and, when war was declared in 1939, the situation was made worse as more vessels were called home. The destroyers of the 21st Flotilla had all left by September 1939. *Eagle*, *Cornwall* and *Dorsetshire* formed a hunting group searching for the *Graf Spee* and the submarines were all withdrawn. These ships were replaced by the new cruiser *Liverpool* and the old cruisers *Danae*, *Durban* and *Dauntless*, as well as some AMCs (Armed Merchant Cruisers), all busily employed in stopping any German merchant ships sailing to the Soviet Union or Europe and searching for contraband. When *Liverpool* stopped the *Asama Maru* and removed some German technicians on their way to Vladivostock, the Japanese government was furious, but had to climb down and agree to refuse passage to German technicians and reservists.

Nevertheless, at the start of the Second World War in Europe, on paper at least, the odds for a war at sea were well in the Allies' favour, with the Royal Navy's strength of 12 battleships, three battle cruisers and six aircraft carriers, added to the French Navy's strength of seven battleships and one aircraft carrier, far exceeding the opposing five German capital ships and, from 1940, the six Italian battleships, a numerical balance which could still have allowed a fleet to be sent eastwards. Qualitatively, the odds were not quite as impressive, as about half of the Royal Navy's strength was old and unmodernized. And between September 1939 and November 1941 the odds were reduced dramatically as the British lost three battleships and three aircraft carriers, and all of the French ships were demilitarized or destroyed following the fall of France and the establishment of the Vichy France state. In the same period, two new battleships and three new aircraft carriers joined the fleet, but the formation and despatch of a Far Eastern fleet became problematical.

One other effect of the fall of France was the Japanese seizure of bases in French Indo-China in April 1941, which allowed Japan the use of air and naval bases only 450 miles from Malaya and 750 from Singapore, well within striking range. The Singapore Defence Conference, held in October 1940, estimated that the Japanese had around 400 land-based aircraft and 280 carrier-borne aircraft that they could use against Singapore, far more than the RAF had in theatre. There were now serious doubts about the speed with which Singapore could be reinforced and with the Period Before Relief now at between 70 and 90 days the Committee asked for 582 aircraft to bolster Singapore's air defence and striking power. Other priorities, principally the necessity of maintaining a strong air defence of Great Britain and of using resources to build up a sufficient force for Bomber Command, as well as the need to re-supply the forces in the Western Desert and the Mediterranean theatre and, from mid-1941, to send aircraft to Russia, all meant that the aircraft required by the command in the Far East were not forthcoming.[4] Of the aircraft in Malaya, most were judged to be adequate. The principal fighter, the Brewster *Buffalo*, was mistakenly reckoned to be comparable to the Hawker *Hurricane*, and of the other aircraft, only the Vickers *Wildebeest* was seen as a second-line aircraft. The *Blenheims*, *Hudsons* and *Catalinas* were all in front-line service elsewhere and judged capable of matching their Japanese counterparts.

By 1940, the China Squadron, comprised of the *Emerald* and the *Enterprise* from the East Indies and the three *D*-class ships, was operating out of Singapore, as Hong Kong was too vulnerable to Japanese attack. Only the three old destroyers, *Scout*, *Thracian* and *Thanet*, four gunboats and the 2nd MTB Flotilla remained in Hong Kong. Britain was now actively seeking US support in the region. In March 1941, the British and US governments concluded the 'ABC-1 Staff Agreement', which provided for Anglo-American cooperation short of war as well as if the United States was drawn into the war. This agreement served as the joint US Army–Navy plan, Rainbow Five, which had the defeat of Germany and its allies as the main, strategic priority, and,

> If Japan does enter the war, the military strategy in the Far East will be defensive. The United States does not intend to add to its present military strength in the Far East but will employ the United States Pacific Fleet offensively in the manner best calculated to weaken Japanese economic power and to support the defense of the Malay Barrier by diverting Japanese strength away from Malaya.[5]

This seemed to be an unequivocal statement of US intentions; the major war effort would be in Europe and the US Pacific Fleet would not be based at Singapore, but at Pearl Harbor, the only other suitable base. Pacific Fleet cruisers might operate from both Singapore and the Philippines, but the main US efforts would be designed to draw Japanese naval forces away from Malaya and the Indian Ocean by operations in the central Pacific, and possibly even by carrier-borne raids against the Japanese mainland.

When Churchill and President Roosevelt met on board the battleship *Prince of Wales* for the Atlantic Conference at Argentia Bay in August 1941, Roosevelt informed Churchill of the extension of the trade embargo as a protest against Japan's actions in mainland China and French Indo-China and the information was immediately signalled to London as a warning that the situation in the Far East was likely to deteriorate. The Joint Planning Staff started to prepare the Far Eastern war plans based on the latest versions of War Memorandum (Eastern), and passed on their proposals to the Naval Staff. Their plan reflected their reluctance to let any modern vessels go eastwards when German and Italian battleships were still afloat and a potent menace to British ships. Even so, the Admiralty's plans remained optimistic, proposing that the Eastern Fleet be composed of the battleships *Nelson* and *Rodney*, the battle cruiser *Renown*, the slow battleships *Revenge*, *Royal Sovereign*, *Resolution* and *Ramillies*, and an aircraft carrier, along with ten cruisers and 24 destroyers. It was hoped that this fleet, together with the presence of the US Pacific Fleet at Pearl Harbor, would act as a formidable deterrent to the Japanese.[6]

This was a fleet designed to impress the Japanese with numbers; quantity not quality was the keynote here, and little or no notice was taken of earlier assessments of the relative fighting qualities of the British and Japanese battleships. Perhaps it would be fairer to say that no notice *could* be taken of earlier assessments, and that numbers were all that the Royal Navy had at this time. The *R*-class ships were slow and not sufficiently armoured to stand up to a concerted air attack, but the more modern aircraft carriers and battleships of the *King George V* class were needed in the Home Fleet to watch the German naval threats to Atlantic and Russian convoys, unless the Americans could be persuaded to provide more naval vessels, including modern battleships, on their Neutrality Patrols and for screening convoys. The Italian Navy also still had to be considered as a fighting force, to be covered by *Queen Elizabeth*-class battleships.

Even so, to repair and refit the vessels for service in the Far East, and to gather sufficient destroyers together from the Home and Mediterranean Fleets, would take time, and the earliest this fleet could be ready was March 1942. This was not purely a military matter, however; political considerations played a part in the decision-making. Churchill disagreed with the planners and maintained that it was possible to send a deterrent force of the newly completed, and still working up, *Duke of York*, with either *Repulse* or *Renown*, an aircraft carrier and supporting ships – echoing his 1925 views on the Japanese when, as Chancellor of the Exchequer, he again expressed his belief that the Japanese would not act offensively against British naval forces, or against Singapore. He believed that all that was necessary to show that the British 'meant business' was a deterrent force operating between Aden, Singapore and Simonstown in South Africa. This would, he minuted, be enough to 'exert a paralysing effect upon Japanese Naval action'.[7]

Churchill saw a smaller fleet of more modern vessels as an economical answer to the dilemma of sending naval forces to Singapore, believing, as he did, that they would have a decisive deterrent effect on the Japanese. In fact, it was hoped that the Japanese, with potentially longer lines of communication between Japan and the Indian Ocean and Singapore, would be put off any naval offensives of their own at all, but that, if not, it would only be with cruisers, armed merchant cruisers or battle cruisers, which the British squadron would be well able to deal with. The loss of the *Bismarck* and the bomb damage to the *Gneisenau* at Brest had altered the naval balance of power in the Atlantic in Britain's favour and, as a result of the Placentia Bay meeting between Roosevelt and Churchill in August 1941, the US Atlantic Fleet took over escorting all convoys to the Mid Ocean Meeting point, just south of Iceland. The pressure in the Atlantic seemed to be temporarily easing on the Royal Navy.[8]

Events were not so favourable in the Mediterranean. Although 1941 had started well with the Battle of Matapan, by the end of May, Greece and Crete had fallen and naval losses had mounted – the aircraft carrier *Illustrious* had been damaged in January, and *Formidable* and two battleships damaged in the battles around Crete. A further three cruisers and six destroyers were sunk and five cruisers and seven destroyers damaged in evacuating the army from Crete. Thus,

> In the course of a year the Royal Navy had lost one battleship sunk and four damaged, one carrier sunk and two damaged, seven cruisers sunk and ten damaged, sixteen destroyers sunk and twelve damaged, one monitor sunk and five submarines sunk and three damaged.[9]

So, when Churchill began to ask for a fleet for the Far East in August 1941, the Admiralty believed that the Royal Navy, and especially its modern battleships, was overstretched. The only battleships in the Home Fleet were the *King George V*, *Prince of Wales* and the recently completed *Duke of York*. *Warspite* was in Bremerton, in the United States being repaired, whilst in the Mediterranean Admiral Cunningham had the modernized *Queen Elizabeth* and *Valiant*. Force H, at Gibraltar, comprised the *Nelson* and *Renown*. Refitting in Britain were *Malaya*, *Repulse* and *Royal Sovereign*, whilst *Rodney* and *Resolution* were refitting in the United States. This left only the unmodernized *Ramillies* and *Revenge*; and they were only suitable for convoy protection.[10] The Chiefs of Staff, debating Churchill's request, agreed that if naval forces were to be sent to Singapore they could comprise one battleship, to be sent from the Mediterranean Fleet, with all four of the *R*-class ships following by the end of the year. The purpose of this force was defensive, not to fight the Japanese but to defend British trade in the Indian Ocean.

Churchill's choice of the third, recently completed, modern battleship, still working up and dealing with constructional defects, the *Duke of York*, would leave *King George V*, *Prince of Wales*, *Nelson* and, once refitted, *Rodney*, supported by aircraft carriers, to deal with any German threat, whilst *Warspite* (once refitted), *Valiant*, *Queen Elizabeth* and attendant aircraft carrier would be sufficient to deal with any Italian moves, leaving the older battleships as convoy escorts and reserves. And Churchill believed that, even if not at war with Japan, the existence of this force and the US Pacific Fleet, with its potential of acting against Japanese Pacific possessions, would be occupying the attention of a major portion of the Japanese Navy. Faced with this as a possibility, the Japanese would not have the major ships to act in force against Singapore and the Indian Ocean trade, but would have to raid with cruisers and armed merchant raiders: 'Churchill therefore wanted a fast *offensive* or "hunting-down" *force* on the lines of that which had so recently destroyed the *Bismarck*. The Admiralty wanted a larger, defensive force of ships to protect Malaya and Singapore.'[11] Like so many others, however, Churchill greatly underestimated Japanese intentions. They were already planning for an attack on Malaya and Singapore, and even if they had intended to send ships to raid British shipping in the Indian Ocean, it is questionable whether the British ships would have been effective against anything bigger than one Japanese battle cruiser. Even if the Japanese raiding force was composed of aircraft carriers, as happened in 1942, there would be little likelihood of the British force surviving long.

The Admiralty were certainly not as confident as Churchill in the ability of the smaller force to exert much influence in the region, either as a deterrent or as a fighting force. When Pound looked at the problem of the Far East in 1940, he had determined that sending an inferior fleet to Singapore was a needless waste of ships, and that they would be better spent operating from Trincomalee, defending convoys. Hence his preference for sending the R class, until a stronger eastern fleet could be formed, once Italian and German naval power had been neutralized. The US offer to defend convoys in the Atlantic, the sinking of the *Bismarck* and neutralization of *Gneisenau* meant that Pound could add *Nelson* and *Rodney*, but not a *KGV*, to the proposed fleet, along with ten cruisers and 32 destroyers.[12]

The Americans, though, were unhappy that such a large British force be employed on trade defence south of the Malay Barrier and not on acting offensively against the Japanese. Churchill was similarly unimpressed, still believing that if a modern battleship was sent it would have the same influence the German *Kriegsmarine* was exerting on the Home Fleet and Russian convoys. Pound's reply to Churchill, in August 1941, modified his position, conceding that he was prepared to deploy *Nelson*, *Rodney*, *Renown* and *Hermes* as soon as possible, with *Ark Royal* following in April 1942. This fleet was not strong enough to engage the Japanese in battle and if war broke out would have to withdraw on Trincomalee, where they would be supported by the four *R*-class ships. Churchill, convinced that the Japanese would not dare risk an attack on Singapore, stuck to his demands for a small squadron of modern ships and the matter came to a head in October 1941.

The reply from Sir Dudley Pound, First Sea Lord, had been quite clear. He did not want valuable ships and men wasted and he restated that it was the Board's decision that all of the modern battleships should remain in home waters, and repeated the Admiralty's intention of sending a large force of older ships, which he and the Board maintained would be the greater deterrent. Churchill angrily responded the next day, correctly stating that the older ships would be unable to catch, let alone fight, a Japanese battle cruiser, and would be useless against anything bigger than an 8-inch cruiser. To him, the *R*-class battleships were nothing more than floating coffins, unable to act against Japanese battle cruisers.[13]

Churchill envisaged the force operating from Singapore, returning periodically to refit and replenish their supplies. The ships would move constantly between Aden, Simonstown and Singapore to avoid being spotted and bombed in harbour and also to keep Japanese forces on their guard, exerting an influence far greater than their numbers would suggest. Japanese forces would have to be on guard everywhere, in case of an attack or raid by the British ships. Given their own assessments of the effectiveness of the Japanese capital ships, the Admiralty's decision to send a larger, slower fleet could only have been made because they recognized the vital importance of defending the Atlantic convoy routes against German surface raiders by concentrating the modern battleships at Scapa Flow, rather than weakening this force to occupy the Japanese. The slower battleships could just as easily defend Indian Ocean convoys and be ready to concentrate to oppose the Japanese when necessary.

So nearly seven weeks elapsed before the issue was raised again, during which time the Admiralty quietly went ahead with its own plans to send the older ships eastwards. But by the middle of October it was obvious, from US sources and from British intelligence sources, that the situation in the Far East was deteriorating more rapidly than had been anticipated. Cable traffic from Tokyo was increasing, Japanese businessmen abroad were selling up and leaving and Japanese merchant ships were cancelling scheduled voyages and returning to Japan. Intelligence reports from Japan remarked on the fact that Japanese merchant ships were gathering in naval ports and having house flags and neutrality national markings removed. More ominously, naval reservists were being called up. When the avowedly aggressive Tojo became Prime Minister in October, diplomats all around the world felt that war in the Far East was moving closer.

The situation had obviously changed for the worse. At a War Cabinet meeting held on 17 October 1941, one day after Tojo's appointment, the British Foreign Secretary, Anthony Eden, asked if it was possible to re-examine the question of sending naval forces to Singapore. Churchill, supported by Eden and Clement Attlee, the Labour leader in the Coalition government, wanted a modern battleship despatched as a visible deterrent to the Japanese, and a signal to Australia and the Far Eastern colonies that Britain took their defence seriously. But Alexander, the First Lord, still persisted in his objections, pointing out that the situation was dissimilar to that of *Tirpitz*; the British ships it was proposed to send to Singapore were to be a deterrent to Japanese raiders, not a raiding force themselves. Slower battleships, escorting convoys, could fulfil this role just as well as modern ships, and the *R*-class ships had already proved this to be a fact on the North Atlantic convoys, where their presence had deterred even the German battle cruisers from attacking the convoys, for fear of critical damage far from a friendly port. A raider that could not find ships to sink and which could not engage a convoy escorted by a battleship was not an effective raider, and the older ships would be just as effective in this role. He was supported in this by Philips, the VCNS, but they were ignored, and when they met again on the 20 October, Churchill had had enough and was not prepared to listen to the Admiralty's views. He and Eden saw the political value that a modern British battleship deployed in the region would have on the

Japanese, and also on the Americans and Australians. He maintained his belief that Japan would not countenance war with Britain but that, if it did, it would probably escalate and include the United States, and with its naval forces in the region as well as the British squadron, superiority would be assured. Reluctantly, Pound agreed to send the *Prince of Wales* to Cape Town so that the publicity of her arrival there *en route* to Singapore could be gained.[14]

It would take several weeks to gather together the ships that would make up an Eastern Fleet: *Prince of Wales* was at Scapa Flow, having recently returned from operations in the Mediterranean, screening Operation Halberd, a convoy to Malta, although *Repulse* was already in the Indian Ocean, screening convoys. The aircraft carrier which was to accompany them, *Indomitable*, was three days out of Greenock, heading for the West Indies, where she would work up and where the air group would join her from training in the United States. Her grounding outside Kingston Harbour, Jamaica, and subsequent repairs at Norfolk, Virginia, effectively ended the deployment with Phillips's force, and there was no other replacement aircraft carrier available. *Ark Royal* had recently been sunk, *Illustrious* and *Formidable* were under repair in the United States following serious damage, *Victorious* was with the Home Fleet watching *Tirpitz*, *Furious* was undergoing a major refit in Britain and was not really a suitable ship, and *Eagle*, *Hermes* and *Argus* were too old, too small and too slow.[15]

The naval situation changed rapidly for the worse in the closing months of 1941, making the debates about the type of fleet to send to Singapore even more heated. As well as *Ark Royal*, the battleship *Barham* was also sunk in the Mediterranean in November 1941, and the battleships *Queen Elizabeth* and *Valiant* were disabled in December. On top of the losses from the evacuation of Crete, this meant that the Mediterranean Fleet was temporarily reduced to a strength of three light cruisers, whilst Force H at Gibraltar was reduced to the unmodernized battleship *Malaya*, the obselete aircraft carrier *Argus* and one cruiser. Not only was the Mediterranean stripped bare of ships, but they were the very ships which could have been of most use in the Far East. The situation regarding destroyers was not much brighter. Only two destroyers, *Express* and *Electra*, were detached from the Home Fleet, and two more, *Encounter* and *Jupiter*, from the Mediterranean Fleet. Both had defects – no admiral would willingly lose his best ships if he could help it. All were sailing to join the *Prince of Wales*, *Repulse* and the destroyers and cruisers of the China Fleet, under Vice-Admiral Layton, who would be in command until the arrival of Rear Admiral Phillips.

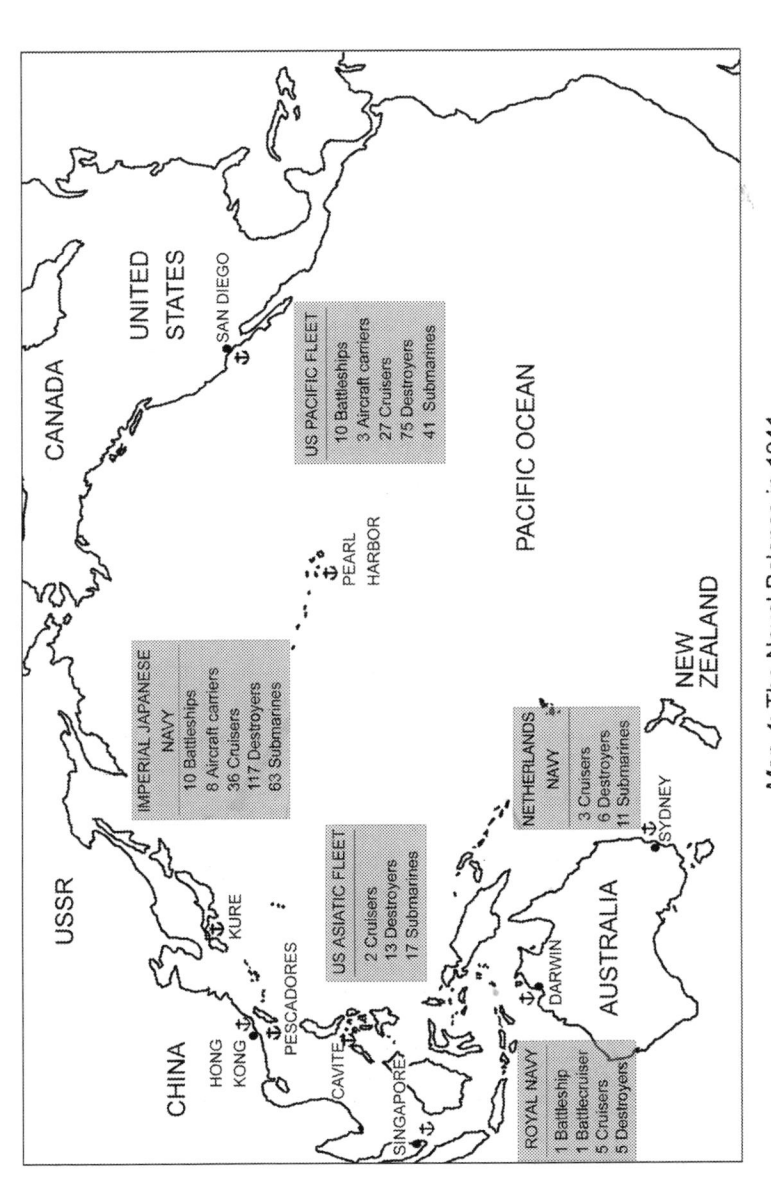

Map 4: The Naval Balance in 1941

Any hopes that these ships would act as a deterrent were quickly dashed when, a week after the War Cabinet's decision to send the ships to Singapore, Naval Intelligence was reporting that the Imperial Japanese Navy was fully mobilized on a war footing and was concentrated at Sasebo naval base in southern Japan.[16] Despite this news, the *Prince of Wales* and her destroyers left the Clyde at the end of October 1941, reached Cape Town on 16 November and rendezvoused at sea with *Repulse* on 29 November. The two ships reached Singapore on 2 December 1941.[17]

Even though the Japanese attack on Pearl Harbor and the invasion of Malaya were only days away, and unsuspected, morale was high in Singapore, bolstered by the arrival of this naval force before a shot had even been fired. On paper there was now a significant naval force at Singapore, prompting Major Fielding Eliot, the *Malaya Tribune*'s military correspondent, to write on the day after the ships arrived at Singapore, joining the old cruisers *Danae*, *Dragon*, *Durban* and the more modern *Mauritius*, undergoing a refit along with the old destroyers *Tenedos* and HMAS *Vampire*:

> A Far Eastern detachment of the size suggested would not be able to seek out the Japanese Navy in Japanese waters and force it to battle, but neither would the Japanese Navy dare venture into the South China Sea ... In fact the arrival of some British battleships at Singapore would render the Japanese naval problem in the Pacific quite hopeless.[18]

The reality was different and, far from being in a hopeless position, the Japanese had already made the decision to go to war as soon after 1 December as possible. They launched no less than five separate operations, against Pearl Harbor, the Philippines, Guam, Wake Island and the Gilbert Islands, Hong Kong and Siam, Malaya and Singapore, all planned for on or immediately after 8 December 1941.

The only worry the Japanese had was that their forces would be spotted and attacked before they could launch their operations, and, in fact, the warships involved in the invasion of Malaya were spotted as early as 21 November; the main units, the old battleships *Kongo* and *Haruna*, were fast, but neither was regarded by the Japanese as a match for the British ships. There were no aircraft carriers with the invasion fleet, but they were protected by two lines of mines laid between the islands of Tioman and Anamba and ten submarines in three patrol lines, with two more boats outside Singapore. Air support was provided by the 22nd Air Flotilla of 36 Navy Type 96 G3M2 *Nell* bombers, which were flown to Saigon, and another 36 went to Tu Duam, north of Saigon. A further 36 fighters and six reconnaissance machines were at Soc Trang, south of Saigon. A further 27 Type 1 G4M1 *Betty* torpedo bombers from the 21st Air Flotilla were also at Saigon.[19]

For their part, the Admiralty were already feeling that the presence of Phillips's force could provoke the Japanese into action and suggested to Phillips that his force should leave Singapore, to disconcert the Japanese who would not know its whereabouts, as well as decreasing the risk to his own ships. On 3 December, they again advised Phillips that Force Z, as the ships were now known, should leave Singapore, possibly in view of the submarine threat – a US patrol aircraft had spotted 12 Japanese submarines heading south off Indo-China the previous day – and advised him to request the use of eight US destroyers that were currently in the Dutch East Indies from Admiral Hart, the commander of the United States Asiatic Fleet. In reply, Phillips informed London that his force was dispersing and that *Repulse* and two destroyers, *Vampire* and *Tenedos*, were to visit Darwin. Phillips hoped to persuade the Australians to attach their light cruiser *Hobart* to his fleet to replace the promised *Sydney*, recently lost (sunk by the German raider *Kormoran*), and as he was going to confer with Hart in any case, he would ask about the destroyers. He must have also had some misgivings concerning the effectiveness of his force, as he requested that the four *R*-class battleships be sent to join his fleet, and for the *Warspite* to call in at Singapore on the way back to Europe from repairs at Bremerton.[20] Such a reinforcement would allow him to increase greatly the scope of his operations, freeing his modern ships for more offensive operations and adding *Warspite*'s 15-inch guns to his force, leaving convoys to be screened by the *R*-class battleships. This was, after all, exactly the plan he had backed just two months previously as VCNS.

The following day, Phillips flew to Manila in the Philippines to meet with Admiral Hart and General MacArthur, to discuss the possibility of cooperation, including the use of both Singapore and Manila by warships of both nations. US warships using Singapore, even if it was just units of the Asiatic Fleet, would further increase Phillips's options. Hart agreed to strengthen Phillips's overall strength by sending four destroyers, the *Whipple*, *John D. Edwards*, *Edsall* and *Alden*, to Singapore, whilst Phillips agreed to move the *Thanet* and *Scout* from Hong Kong to Singapore. If he had received all of the forces that he was requesting, he would have commanded a force of one modern and one modernized battleship, a battle cruiser, between four and six cruisers, and 12 destroyers.

On the same day, 6 December, a convoy of 25 Japanese merchant ships, escorted by a battleship (actually a cruiser), five cruisers and seven destroyers, was sighted by RAAF *Hudson* well south of Saigon, sailing west. In response to this obvious threat the four US destroyers sailed from Balikapan in the Dutch East Indies, whilst *Repulse* and its destroyers were recalled to Singapore.

Three more convoys were also reported, but Japanese luck held. The countries were not yet at war and not only were all the convoys well beyond the range of any British aircraft, but the British naval forces were dispersed and the Flag Officer, Phillips, was in Manila. Not until 7 December were Force Z and their admiral reunited, when both Phillips and *Repulse* arrived back in Singapore. Other vessels were also steaming to join his flag at Singapore, principally the cruiser *Exeter*, escorting a convoy across the Bay of Bengal, which was ordered to detach and sail immediately to join Phillips.

Just after midnight on 7 December the first Japanese troops landed near Kota Bharu in Malay and overwhelmed the defenders. The nearby airfield was abandoned when ground staff panicked and retreated. Unlike the pre-war exercises, the Japanese had possession of an airfield within hours of landing, not weeks. On the same day, Japanese aircraft attacked the US naval base at Pearl Harbor and sank or crippled all eight US battleships of the Pacific Fleet, and destroyed 188 aircraft, whilst Hong Kong, the Philippines and Singapore were also attacked by aircraft. The war in the Pacific had begun, and both the Americans and British had suffered their first reversals of fortune. US naval and air power had been eliminated at a stroke, and only Phillips's force of two capital ships, supported by cruisers, destroyers and minor vessels, could now act against the Japanese. The deterrent force had suddenly become the principal Allied naval presence in the Pacific and Japan's main naval target.

With heavy fighting in the north of Malaya, a conference was held on *Prince of Wales* on the day after these attacks to determine the naval response to the situation. The capital ships of Force Z, *Prince of Wales* and *Repulse* were ready for sea, but of the cruisers, only the old light cruiser *Durban* was ready for sea. Of the modern cruisers, *Mauritius* was undergoing a refit, whilst *Exeter* and the Dutch *Java* would not arrive for another 36 hours. Of the destroyers, *Jupiter* and *Encounter* were repairing defects, but *Express*, *Electra*, *Vampire*, *Tenedos* and *Stronghold* were ready for sea, and in two more days the four US destroyers and two British destroyers from Hong Kong were expected. On balance though, it appeared that Phillips could still take a formidable force of one modern battleship, one battle cruiser, one light cruiser and five destroyers to sea to attack Japanese transports, and he decided to proceed without *Durban* or *Stronghold*. Force Z sailed at dusk, for Singora.

Unluckily for Phillips, his ships were sighted moving north by a Japanese submarine, and the battleships *Kongo, Haruna*, four cruisers and several destroyers were ordered to concentrate, ready for a surface action, whilst the transports were ordered to scatter to the north.[21] Japanese floatplanes from their naval forces sighted Force Z in the late afternoon, but Phillips narrowly avoided a surface action with the Japanese after altering course to investigate reports of further landings at Kota Bharu. Believing his mission to be compromised by the sighting, he cancelled his operation and his ships turned back to Singapore. On 10 December he received reports of more landings at Kuantan, and Phillips again altered course to investigate these reports. In doing so he allowed himself to be spotted again and reported, first by the submarine *I58* and then by aircraft.

A *Walrus* flying boat which had been catapulted off by Phillips found no sign of landings at Kuantan, but by now it was too late, and shortly after 11.00 a.m. his force was attacked by Japanese planes, which launched a series of bombing and torpedo attacks, sinking both ships.[22] Air cover had been belatedly called for, but when the Brewster *Buffalo* fighters arrived, they were too late to do more than fly air cover over the destroyers as they rescued survivors.

The sinking of *Prince of Wales* and *Repulse* was a triumph for Japanese naval aviation. For the first time ever, capital ships under way had been attacked and sunk by aircraft, unaided by any surface vessels. Even fighter cover would not have made any appreciable difference to the outcome. Some Japanese aircraft may well have been shot down, but the Japanese airmen had been trained to fight their way through to the targets, and the Brewster *Buffalo* fighters would soon have been overwhelmed.

Allied naval power in the Pacific had been further reduced to scratch forces of Australian, British, Dutch and US cruisers and destroyers, and Admiral Somerville's hastily gathered Eastern Fleet in the Indian Ocean. There was to be no 'Main Fleet', and no 'Three-Phase War Plan'. War Memorandum (Eastern) went down with the *Prince of Wales* and *Repulse*, whilst the Battle of the Java Sea saw the Allied cruiser force destroyed in February 1942. The Japanese took Malaya, Singapore and the Dutch East Indies and when Admiral Somerville was faced with the prospect of combat with Admiral Nagumo's aircraft carrier force in the Indian Ocean in early 1942, his force, including the modernized battleship, *Warspite*, and the *R*-class battleships, could do little. The *R*-class ships had all had three years of arduous war service and were old, slow vessels of limited fighting value, and not fit to face enemy battleships, even in a night action. As he wrote to Pound, on 12 February 1942: 'The *R*-class battleships are old ships of relatively low fighting value and have in consequence been relegated mainly to convoy escort duty, a duty which provides small opportunity to promote fighting efficiency.'[23] Whilst to his wife, he wrote:

> I hear a lot of blah about how everything now depends on our maintaining control of the Indian Ocean. That's poor bloody me and I wonder how it's to be accomplished. My old battle boats are in various states of disrepair and there's not a ship at present that approaches what I would call a proper standard of efficiency.[24]

Somerville knew that it would have been suicide for him to engage the enemy with a fleet of old, unsuitable vessels which had not had the opportunity to work together as a battle fleet, or for their captains to meet with, and confer with, their admiral. His Eastern Fleet was a fleet in name only, despite the presence of two of Britain's latest aircraft carriers. Somerville's role was to keep his fleet afloat, to keep a naval presence in the region, whilst avoiding combat. His was a 'fleet-in-being', and pitting his ships and aircraft against the Japanese defences would have been suicide. More tellingly, in April, he wrote to his wife, 'until I get a proper fleet out here I will simply have to hide'.[25] And that is exactly what he did, and in so doing he preserved Britain's tenuous hold on the Indian Ocean. Somerville had no alternative. With the Japanese aircraft carrier force of *Akagi*, *Soryu*, *Hiryu*, *Zuikaku* and *Shokaku* screened by the four battleships, *Kongo*, *Kirishima*, *Hiei* and *Haruna*, at large in the Indian Ocean and the light carrier *Ryujo* and five heavy cruisers in the Bay of Bengal, the preservation of Somerville's force, the battleships *Warspite*, *Royal Sovereign*, *Resolution*, *Ramillies*, *Revenge*, the aircraft carriers *Indomitable*, *Formidable* and *Hermes* and the cruisers *Dorsetshire*, *Cornwall*, *Emerald*, *Enterprise*, *Caledon*, *Dragon* and the Dutch *Jacob van Heemskerck*, inferior as it was, as a 'fleet in being' was vital. This was an inadequate naval force with which to engage the Japanese, and Somerville knew it, hampered as he was by the lack of adequate air cover and with the unsuitable *R*-class battleships. But his ships represented the only alternative to handing the Japanese complete and undisputed naval superiority in the Pacific and Indian Oceans. As it was, he lost *Hermes*, *Cornwall* and *Dorsetshire* to Japanese air attacks, and his other ships narrowly missed being detected by Japanese aircraft.[26] However, Somerville's ships survived and his fleet became the nucleus of a more powerful Eastern Fleet, supporting later operations in Burma and Malaya and allowing for a British naval presence in the Pacific during 1945.

NOTES

1. Brice, *The Royal Navy*, p. 45.
2. Ibid., pp. 54–5.
3. Ibid., pp. 121–31.
4. Middlebrook and Mahoney, *Battleship*, p. 21.
5. E. King and W. Muir Whitehead, *Fleet Admiral King: A Naval Record* (Eyre & Spottiswoode, London, 1953), p. 121.
6. Middlebrook and Mahoney, *Battleship*, p. 27.
7. Ibid., p. 29.
8. R. Brodhurst, *Churchill's Anchor: The Biography of Admiral of the Fleet Sir Dudley Pound, OM, GCB, GCVO* (Leo Cooper, London, 2000), pp. 184–90.
9. Ibid., p. 191.
10. Ibid., p. 193.
11. Middlebrook and Mahoney, *Battleship*, p. 30.
12. Brodhurst, *Churchill's Anchor*, p. 194.
13. Middlebrook and Mahoney, *Battleship*, p. 31.
14. Ibid.; CAB 69/2 69/8, pp. 33–5.
15. Brodhurst, *Churchill's Anchor*, p. 197.
16. Marder, *Old Friends, New Enemies*, p. 215.
17. Middlebrook and Mahoney, *Battleship*, pp. 62–73.
18. Ibid., p. 75.
19. Ibid., pp. 83–4; Peattie, *Sunburst*, pp. 298–301, for details of these aircraft.
20. Middlebrook and Mahoney, *Battleship*, p. 88.
21. Ibid., p. 129.
22. Ibid., pp. 288–91.
23. M. Simpson (ed.), *The Somerville Papers: Selections from the Private and Official Correspondence of Admiral of the Fleet Sir James Somerville, GCB, GBE, DSO* (Navy Records Society/Scolar Press, Aldershot, 1995), p. 384.
24. Ibid., p. 394.
25. Ibid., p. 401.
26. I. Ballantyne, *Warspite* (Leo Cooper, London, 2001), pp. 149–50.

Conclusion: War Memorandum (Eastern) and the Royal Navy's Strategic, Operational and Tactical Development

> A classic, proactive method of achieving sea control frequently advocated and used by the Royal Navy in the past ... is to seek out the enemy to bring him to decisive battle, thereby destroying his forces and eliminating his capability to challenge sea control. Other maritime operations can then proceed unthreatened. Historically this course was not simply a matter of élan but reflected Britain's qualitative superiority over its enemies.[1]

In both its first form and in later re-writes, the British plan for fighting the Japanese in the Far East, War Memorandum (Eastern) was a strategy that responded to these concerns; initially as an aim in itself, and then, later, as a means of protecting British cruisers attacking Japanese trade, the Japanese fleet was to be sought out and destroyed in a decisive battle. Areas of ocean could not be occupied in the way that areas of land could be; they had to be controlled through the exercise of strategic maritime power and a powerful battle fleet was widely recognized as the way of achieving this during the 1920s. The presence of such a fleet in an area did not on its own constitute a naval strategy; it had to conduct operations which would stop an enemy from using the seas without great risk to itself.[2] In the Admiralty's Far Eastern war planning between 1919 and 1931 this was to be the destruction or neutralization of the Japanese battle fleet, so that, in the aftermath of the Washington Conference and the ending of the Anglo-Japanese Alliance, the Royal Navy succeeded in identifying a possible enemy, maintained a powerful battle fleet capable of meeting them in battle and refined the strategy and tactics with which to defeat the enemy or deny them the free, uninterrupted use of the seas and oceans.

The fact that this would have been the first occasion when the battle fleet would operate thousands of miles from home waters made the strategy appear to be logistically impossible, if not unrealistic. But the Admiralty justified such long-term planning; navies take decades to develop, and large warships cannot be built when a threat appears. What became important was to have a strategy which identified an enemy and, through this, the ships needed to fight them. Thus the Admiralty had to have a strategy which justified having a large battle fleet and the high budgetary expenditure to maintain such a fleet. Only in this way would the Admiralty feel confident enough to carry out its various imperial duties.

At the end of the war, superficially at least, relations between the Royal Navy and the Imperial Japanese Navy remained good, and when Jellicoe, in his submission to the Admiralty on imperial naval policy, re-emphasized Dominion concerns about Japan, his suggestions were rejected and he was told he had exceeded his instructions.[3] But although the size of the fleet he proposed for the Far East was unrealistic to many within the Admiralty, including Beatty, Japan seemed to offer the only real threat to the Empire, a threat which had to be guarded against. And the best and most effective way of doing this was by having a large battle fleet, able to deploy eastwards in times of tension, and the logistical infrastructure to support this fleet and a fully developed, large naval base, able to act as the base for this fleet and its operations.

Strategists have always been fascinated by the theory of the quick war and decisive battle, and the Plans Division was no exception. The first plans for a war in the Far East were simply plans to move ships into Singapore. Just what the fleet would do once in Singapore, beyond fighting the Japanese, was left vague and up to the man on the spot. This was not a naval strategy, and it placed too much dependence on the idea of a decisive naval battle resolving the war, when recent experience had clearly shown this would not be the case. Once the naval strategists started to refine these plans through exercises and analysis, they began to realize that a quick victory was less likely than a long war of attrition, with the disruption of Japanese maritime trade as its basis.[4] British naval planners assumed that Japan would be vulnerable to economic warfare, and expected Japan to accept British terms if the Royal Navy severed the sea lines of communication.[5]

Achieving this would entail deploying significant naval forces, most probably the whole of the Royal Navy's battle fleet, in the Pacific region for an unspecified amount of time. What therefore became an important consideration was the close link between the strategy and the size of the battle fleet, and the extent to which the war planning was used, either explicitly or implicitly, as a means of justifying a large battle fleet, regardless of the strategic threat. Japan certainly remained the main strategic threat as far as the Admiralty was concerned, despite advice to the contrary by the Foreign Office, which identified the main threat to British interests in the Far East as arising from the spreading of Bolshevik influences in China and from the nationalist Kuomintang.[6] The Foreign Office wanted good relations with Japan, the Admiralty a large battle fleet.

In reality, the Admiralty did not develop one Far Eastern strategy, but several. The initial plans, between 1919 and 1931, including the first versions of War Memorandum (Eastern), envisaged sending out a superior battle fleet to fight the Japanese, whereas by 1939 it was intended that the fleet being sent would be inferior to the Imperial Japanese Navy, relying on qualitative superiority, long Japanese lines of communication and the United States to provide the advantage. It was no longer regarded as probable that the battle fleet would be able to proceed north from Singapore without US help, although,

> If Singapore could have been held, Britain might have produced a stalemate which left most of its vital interests intact. With American support and the mobilization of imperial resources the Royal Navy might eventually have adopted a more aggressive strategy.[7]

Of necessity, this would have been a long-drawn-out war, and by the time that Churchill wanted the Admiralty to commit forces to the region, they could no longer hope to provide a battle fleet of modern warships, being heavily committed in the Mediterranean theatre, the North Atlantic and the Russian convoy routes. This, plus war losses, meant that the forces necessary to hold Singapore were just not available, and the alternative, the combination of the American, Dutch, Free French, Australian and British vessels, was all but destroyed in a series of actions starting with the attack on Pearl Harbor and concluding with Nagumo's raids in the Indian Ocean. Until the United States managed to inflict a crushing blow against the Japanese at Midway, allied naval power in the Pacific was forced onto the defensive.

An important consideration when developing the first versions of War Memorandum (Eastern) was the continued provision of a superior battle fleet to allow a British deployment to the Far East and the Japanese fleet was the only justification for a large Royal Navy that the Admiralty had; a battle fleet sent to the Far East would need a clear, quantitative and qualitative superiority over the Japanese fleet. The more ships the Japanese built, the more ships the Royal Navy would need to have. That was, in many ways, the easy part. Training commanders and men to respond to new conditions of warfare would take longer, and its success would be more imponderable. Only war would reveal the success of training.

CONCLUSION

The deliberations about the future of the capital ship in the Royal Navy through the Post War Questions Committee and the Bonar Law Enquiry had confirmed the primacy of the battleship before the Washington Conference endorsed this for the world's navies. The Washington limits and ratios had established a naval balance in capital ships which left the British with a battle fleet equal to the US fleet and larger than the Japanese. Somewhat strangely, this, in its turn, prompted the Admiralty to stress the Japanese naval threat, especially with regard to cruisers, however realistic or otherwise it was, starting what was, in effect, a new naval race, but this time in cruisers.[8] But to many naval officers this made sense. They were suspicious of Japanese naval ambitions and saw Japan not as an old and trusted friend, but as a potential, in fact for many years, the only, enemy, and the Royal Navy had to be ready and able to deal with it.

As a consequence, the Admiralty proposed a British naval building programme as a response to the 1922 Japanese announcement that the Imperial Japanese Navy planned to replace most of its lighter warships and build new cruisers. This Japanese programme was certainly exaggerated by the Admiralty to justify the claim that only by developing a Far Eastern strategy, based on sending the battle fleet to Singapore, could any balance in the region be achieved, since, without a fully functioning, major fleet base, it would be impossible for a British battle fleet to operate in the Pacific if relations with Japan ever cooled in the future. Interpreting the Ten Year Rule as meaning that all naval preparations were to be ready by 1929, the Admiralty obstinately

> used that rule to justify increasingly large expenditure on its programmes. It also told the Cabinet that if any of its programmes were delayed the government must publicly abandon the one power standard. Then, the Admiralty used the Japanese menace to justify the development of two means to project great offensive power to every sea. The first was to complete the fuel oil reserve and the Singapore base for the fleet to operate in the Pacific Ocean, off the coast of Malaya – or of Japan. The second was to maintain the world's greatest fleet.[9]

But even though it managed to maintain the world's largest fleet and to stockpile 60 per cent of the necessary oil fuel reserves, the Admiralty recognized an unreality in the arguments, shrinking, as it did, from the financial challenges of developing a base which could cope with a fleet of 15 battleships – and all of the cruisers and destroyers which would accompany them – in favour of the 'Red' scheme, which planned a base that would be able to cope with four battleships, and which would have a graving dock and a floating dock, 6,000 feet of wharves and ancillary facilities.[10] But of course, once built, Singapore could be expanded and if the strategy developed as planned, the bulk of the battle fleet would be operating from Hong Kong or an advanced base, not Singapore. Even such a strategy demanded a large battle fleet and, as a consequence, a large Royal Navy. Anything less than a fleet superior to the Japanese battle fleet, able to relieve Singapore and move northwards, would be no deterrent to the Japanese and might have the opposite effect, offering up a tempting target to them.[11] A large battle fleet and a Far Eastern strategy went hand in hand; one was necessary for the other to be viable. And yet, despite the apparent emphasis on the battleship during the inter-war period, the Royal Navy actually spent less money on modernizing its battle fleet than either of its chief rivals, the United States Navy and the Imperial Japanese Navy, and instead put more emphasis on a six-year programme to build 70 cruisers, of which no more than ten were to be over 15 years old, aircraft carriers to replace the small *Argus*, *Eagle* and *Hermes*, destroyers, submarines and auxiliaries, and only proposing to build seven new battleships in 1931, after the end of the Washington Treaty's ten-year building holiday. This would provide sufficient strength both to defeat Japan and to leave the Royal Navy equal in strength to the United States Navy at the end of the war.[12] By skilfully stating its case and lobbying successive governments, the Admiralty had managed to maintain what was probably the strongest navy in the world, with 20 capital ships as compared to the United States' 18 and the Japanese ten, as well as the largest number of cruisers and aircraft carriers.[13]

By way of a contrast to the stated Admiralty concerns about Japan, which were echoed by some Dominion premiers, Churchill, the Chancellor of the Exchequer, and Austen Chamberlain, the Foreign Secretary in 1925, were both quick to dismiss Admiralty fears, believing that if such a threat materialized, then the Royal Navy, as it was, would have no difficulty in sweeping Japan off the seas in three to four years. So, in the same year as the Admiralty was planning its enlarged fleet, and seeking comments on the Japanese threat and its newly developed War Memorandum (Eastern) from commanders-in-chief in the East Indies and China stations,

> the Committee of Imperial Defence intimated that there was no such threat. However, the wording of the statement, 'In existing circumstances Japanese aggression against the British Empire is not a contingency to be apprehended for the next ten years' sounds less a conviction of the non-existence of a Japanese threat than a drastic adjustment of strategy to budgetary realities.[14]

CONCLUSION

From a naval point of view, any cuts in the size of the fleet or the budget would serve neither the interests of the Royal Navy nor the Empire. Any British weaknesses, real or imagined, in the Far East could present Japan with the opportunities it sought to expand its territory. This confirms the likely explanation that

> many seemingly anti-Japanese Navy statements were more motivated by an Admiralty desire for greater budgetary allocations for the Singapore naval base than by any real fear of imminent Japanese aggression against the British Empire. It is, in fact, extremely difficult to clarify whether Japan was the budgetary or potential enemy although Beatty, for example, seems always to have distrusted the Japanese.[15]

Deciding whether the strategy being developed dictated the size of the fleet, or the size of the fleet necessarily dictated the strategy being developed becomes the problem, and hinges on the speculations associated with whether the strategy could have worked. Certainly the plans were not without critics. Rear Admiral Richmond, the Commander-in Chief, East Indies Fleet, believed the 1924 plan, War Memorandum (Eastern), to be unimpressive and unrealistic, just as he believed that the Royal Navy would be better off with smaller battleships of some 10,000 tons and armed with six to eight 12-inch guns.[16] Progress of sorts was made through the combined exercise at Salsette Island in 1925, which seemed to suggest that Singapore would be secure as long as it was capable of defending itself for a period of 35 days, making any Japanese invasion a risk. If this were to be the case, and if the Japanese were not going to accept the risk of moving thousands of miles south with their battle fleet to attack Singapore, the problem then was how to bring them to battle and defeat them. So, War Memorandum (Eastern) evolved into a three-phase war plan, based around attacking Japanese trade as the means to draw out the battle fleet.[17]

For the British this would mean not only being able to deploy to Singapore, but also the retention or recapture of Hong Kong and the capture of an advanced base near Japan. Cruisers and submarines would attack Japanese trade, the battle fleet would attack enemy cruisers and detached battle squadrons and thus would force out the Japanese battle fleet to fight the decisive fleet action. Battleships, battle cruisers, aircraft carriers, cruisers and destroyers would all be needed for the decisive battle, but more cruisers and submarines would also be needed for the strategic role of attacking the Japanese trade routes. Tactically there was also the question of how to use the battle fleet, which was smaller than the Grand Fleet had been, but still one of the world's most powerful fleets. Successfully executing a strategy depended on more than just numbers, and operations against the Japanese conducted at great distances from naval dockyards needed to be meticulously built up and resolved quickly. The decisive battle would have to be fought fiercely and resolved quickly.

The reference point for all of the developments that took place in tactics was, inevitably, Jutland, given that it was the only large clash of dreadnought battleships, although this should never be taken to imply that the Royal Navy (or for that matter the United States Navy, Imperial Japanese Navy or, indeed, any of the world's navies that sought to profit from the lessons to be learnt) was simply interested in re-fighting that battle, in the waters of Japan or anywhere else.

The organization of the Grand Fleet, and how this affected the way the battle was fought, was a crucial aspect of study. The two British commanders, Jellicoe and Beatty, were products of the same system, but were almost exact opposites; Jellicoe was at his best with technicalities, but he was an authoritarian leader, whereas Beatty was a more permissive leader. According to Andrew Gordon, Jellicoe was a 'regulator' whilst Beatty was a 'ratcatcher'. One tried to control all aspects of his fleet through orders and signalling, whilst the other, recognizing that signalling might break down in the heat of battle, and being aware of the chaotic nature of warfare, allowed his subordinate commanders to react to situations.[18]

As First Sea Lord, Beatty wanted to be concerned with the future and not the past; he actively did not want to re-fight the Battle of Jutland, but wanted to prepare the Royal Navy to win the next battle, decisively. Under his stewardship as First Sea Lord, the whole Admiralty climate promoted this, and the tactics of the smaller Royal Navy battle fleet were rewritten to reflect this. Madden's 1921 Battle Instructions were more permissive, a refinement of Beatty's Grand Fleet Battle Instructions; Madden was realistic enough to see the changes were necessary. Drax and Richmond looked at the shortcomings in the way the Battle of Jutland had been fought and saw, in the future, a different naval battle evolving, with aircraft bombing and machine gunning, destroyers darting forward to launch torpedoes, and battleship squadrons acting independently, and at close range, to overwhelm and defeat their opponents. This was one of the reasons that Richmond gave for smaller battleships, the basis of his theory being that numbers and firepower would ensure victory, that a fleet with 15 smaller ships, opening fire as early as possible, would overwhelm and eliminate an enemy fleet of ten.[19] This was a quantitative approach to battle fleets and, whilst it was important, the qualitative aspect, which could be gained only through practice and

CONCLUSION

experience, was equally so. Development of the qualitative side of naval warfare could only come through actual wartime experience, or, in peacetime, through analysis of wartime experiences, exercises and training.

Drax and Richmond both also lectured and wrote on decentralization in battle, and when First Sea Lord in 1938–39, Backhouse, a centralizer with the reputation of being unable to delegate, did bring Drax back from retirement to rewrite the Fighting Instructions. There was to be a measure of independence for subordinate commanders, but the Admiral was still to have control over his battle fleet until circumstances favoured independent action. What was different was the emphasis on the use of initiative, the assumption that subordinates did not have to wait for orders, that the Admiral did not necessarily know everything that was going on, and on the repeated testing of different theories and tactical situations, at fleet, squadron and ship level, to emphasize the importance of using initiative as a part of the fleet. A whole variety of exercises and tactical problems had to be analysed and criticized, as

> the danger inherent in any doctrine is that without careful, critical examination, testing and re-examination it can deteriorate ... into a positive, arrogant assertion of opinion. This ... can become a mental straitjacket for a commander when a specific set of events require flexible thinking.[20]

This was what was widely perceived to have happened at Jutland, and to have contributed to the failure to achieve a decisive result. Naval warfare was not just a question of numbers of ships or guns, but was a synthesis of the material, tradition and training. Officers unaccustomed to constant, rigorous training would be hesitant in battle, and so training had to be as realistic as possible, so that any mistakes would be made in training, not in the decisive battle:

> Thus all processes of battle fleet operations were meticulously exercised, divisional tactics were explored, night fighting skills developed, communications improved and so on ... the navy's instrument would be a balanced fleet comprising a judicious mix of all types of war vessel co-operating in mutual support.[21]

But the Fighting Instructions also made it clear that, whilst subordinates were not to expect a succession of signals or instructions from him, the Admiral was still in overall control, and squadron commanders were expected to both observe what his flagship was doing and follow its lead, and to be familiar with his intentions in battle. By using other publications such as the Naval War Manual and by reading the annual publications on exercises and tactical progress, subordinates could also become familiar with diverse courses of action and could test them out in manoeuvres, discussions and exercises. Thus the naval war games expounded doctrine, made officers aware of the potential of their squadrons, ships, men and weapons, and helped validate tactics and future warship design. As Moretz states:

> The primary objective of Royal Navy tactical thinking was to ensure superior firepower during any fleet to fleet action ... The irony is that from a tactical viewpoint, the one encounter the Royal Navy had consistently trained to meet was the one it was least prepared to deal with when conflict came due to commitments in Home, Atlantic and Mediterranean waters: war with Japan.[22]

The war in the Far East did not happen in the way that the writers of War Memorandum (Eastern) envisaged. It had always been acknowledged that the situation in Europe needed to be stable for the plan to be executed as written, and as events in the 1930s showed, this was not the case. Italian naval expansion and the rise of Nazi Germany both constrained the Royal Navy's ability to act in the way originally envisaged in War Memorandum (Eastern). By 1934, European politics had induced something of a sense of inertia into British foreign policy, with the government responding to international events as they happened.

During the 1930s the 'New Standard' fleet being developed under Chatfield was planned to operate as a battle fleet with attached aircraft carriers and, in the case of operations in the Far East, this would mean carrying more bombers and torpedo bombers to strike against the Japanese fleet, and, importantly, the aircraft carriers, as well as slowing down the enemy battleships, would allow the British battle fleet to draw up and engage in a close-range, night action.

Air raids on warships at anchor were not a feature of the Royal Navy's tactical planning until 1938, when a plan to attack the Italian fleet at Taranto was resurrected. On the face of it, this seemed surprising, as in 1918 an air raid on the German High Seas Fleet at Wilhelmshaven had been planned, using 100 aircraft and four aircraft carriers, and in a 1919 exercise, an aerial attack on Portland Harbour by Sopwith *Cuckoos*, the air-launched torpedo had proved itself to be a decisive weapon.[23] One drawback of this sort of attack was, of course (as proved by the attacks on Taranto and Pearl Harbor), that ships sunk in harbour could be repaired and rejoin the fleet as effective units. This had to be balanced against the psychological

CONCLUSION

effect that such an attack could have, however. Only as the fleet moved into striking distance of the Japanese mainland would more fighters be embarked, to escort the air strikes which would be launched against Japanese warships. Royal Navy aircraft carriers, operating with the battle fleet and screened by cruisers and destroyers and their anti-aircraft batteries, would not carry fighters for fleet defence, relying instead on the combined barrage, and later on their armoured flight decks, to protect them.

The Royal Navy saw roles for naval aviation, spotting and slowing down an enemy, allowing the battleships to finish the enemy off with gunfire (the doctrine of 'Find, Fix and Fight'), and the apparent lack of development of naval aviation in Britain between 1919 and 1939 was as much down to the roles that the Admiralty prescribed for its naval aircraft as it was to RAF intransigence. The Admiralty was happy with low-performance biplanes if they did the job required of them – hence the long, front-line service life of the Fairey *Swordfish* – and, mistakenly, did not see the need for a high-performance fighter aircraft. With the flying expertise in the Fleet Air Arm of the RAF and not in the Royal Navy, if the experts said that high-performance aircraft made deck parks and crash barriers too dangerous, then the Admiralty went along with this, the experts' opinion, and took little notice of developments of these on USS *Langley*.

Fighter interception had been highly regarded in the 1920s, but as the performance of aircraft increased it became more and more difficult to provide enough early warning time to launch an interception strike without pushing the *A-K* cruiser line, which provided the first warning, out beyond a point where it could be supported by the battle fleet. There was a change in the ratio of fighters and bombers carried; one fighter to every two bombers in the 1920s, but in the 1930s, with larger aircraft, it was only one fighter to every five bombers. And as every fighter carried was one less strike aircraft, the few fighters carried by the aircraft carriers of the late 1930s were all low-performance, designed to protect air strikes, not the fleet. More faith was placed (or, perhaps more accurately, misplaced) in the High Angle Control System Mark I, Vickers Predictor and the Mark M multiple pom-pom quick-firing gun, designed to place a barrage of explosive shells in the path of any aircraft attacking the fleet, so that,

> On the outbreak of war in 1939, British warships were equipped with fire control outfits that were incapable of dealing with rapidly moving aircraft, while the low-angle main armament of most destroyers – whose intended targets were other ships – was virtually useless against high flying and dive bombers.[24]

Financial limitations between the First and Second World Wars were undoubtedly a contributory factor to this and despite the increases in the number of close-range anti-aircraft guns carried, including the highly regarded 2pdr multiple pom-pom gun and the new, dual-purpose batteries on all new battleships, aircraft carriers, anti-aircraft cruisers and many destroyers, the anti-aircraft capability of many British warships was less effective than was claimed.[25] Many destroyers had a main battery which was incapable of sufficient elevation to engage aircraft, for example, and relied instead on Lewis guns, 0.5-inch machine guns, 2pdr pom-poms and a single 3-inch or 4-inch high-angle gun, in addition to their high speed and manoeuvrability, for their anti-aircraft defence.

That not every naval officer was as confident in the anti-aircraft arrangements is evidenced by Rear Admiral Henderson, writing in 1932, and warning that

> The primary defence of the fleet against air attack is not justified by any data or experience. No realistic firing against aircraft has taken place since the last war and, in my opinion, the value of our own High Angle Control System Mk1 is rated too high. In common with others we are apt to over-rate the capabilities of our weapons in peacetime.[26]

Whereas most naval officers believed in the superiority of an anti-aircraft defence, Henderson was not as confident, pointing out that during the First World War more aircraft had been destroyed in aerial combat than by anti-aircraft fire. He further believed that the smaller numbers of aircraft employed at sea would make interception easier than over land but, until an effective means of early warning and a fighter with sufficient endurance and firepower could be developed, the Royal Navy would have to continue to rely on high-angled barrage fire.

Two years later, in January 1934, the First Lord of the Admiralty, Sir Bolton Eyres-Monsell, writing to Sir Philip Sassoon, Under-Secretary of the Air Ministry, pointed out the growing gap between the Fleet Air Arm's operating procedures and those of its rivals when he wrote:

> In our view the last fifteen years have fully demonstrated the effectiveness of aircraft carriers as essential components of a modern fleet ... We are fully aware that the operating capacity attributed to some of the smaller US and Japanese carriers is much larger. I understand that American operating capacity includes a larger number of aircraft permanently parked on deck, for which there is no stowage

room in the hangars ... We know less of Japanese than of American methods of operating aircraft from carriers, but here again, it is possible that in order to operate 32 aircraft from a 10,000 ton carrier the Japanese have followed the American practice of stowing aircraft in the hangars with their wings detached ... you will see that even in 1938, when the proposed new carrier is completed, we should be inferior to the Fleet Air Arm of the Japanese Fleet, even if all our carriers could be sent to the East, and this, as I have already explained, could not be done.[27]

By comparison, the Japanese had looked to naval aviation as a way of redressing the imbalance brought about by Washington and so developed a different strategic and tactical doctrine for naval aviation. They wanted planes with a long range, to ensure that they could attack enemy carriers before their own were detected and attacked. And, in any war with the Japanese, the British Main Fleet would be operating thousands of miles away from home waters and replacement aircraft, opposing all of the Japanese carrier-borne, and a fair number of land-based, aircraft with less than the whole Royal Navy's aircraft carrier strength. Even if shore-based aircraft were available, they would have to be operating many hundreds of miles from their main bases. To Monsell, therefore, it was essential that the Fleet Air Arm became an integral part of the Royal Navy, as the fleet would be at a considerable disadvantage if faced by the Japanese fleet without a fleet air arm under naval control. By this time, the Japanese had started to acquire a very different view of aircraft carrier warfare than the British had developed, using their aircraft carriers in a fast, striking squadron, supported by battleships or battle cruisers, unlike the British aircraft carriers, which were still expected to be operating with the battle fleet, providing spotting and reconnaissance aircraft as well as torpedo bombers. And with the Admiralty's insistence on low-performance fighters and multi-role aircraft, it is likely that the Gloster *Sea Gladiator* biplane fighter, the Fairey *Fulmar*, fleet fighter and spotter, the Blackburn *Skua*, fighter/dive bomber, and the Blackburn *Roc*, a low-performance fighter with a quadruple gun turret, would have been significantly outclassed by their Japanese counterparts whilst the biplane torpedo bomber, the Fairey *Swordfish*, would have fared no better in the face of Japanese anti-aircraft fire and interception by the Mitsubishi *Zero* fighters.

From *Ark Royal* onwards, aircraft carriers were well-protected, enclosed floating armoured boxes. The flight deck became the main strength deck, and the Royal Navy's operating practice was to keep all the aircraft below decks, protected by the armoured deck, only ranging the aircraft on deck when they were to be launched, being brought on deck via a rear lift. The belated introduction of a midships crash barrier allowed for some deck parking on the *Ark Royal* and *Illustrious* class, but to nowhere near the extent that the Americans had adopted, until war experience proved the need for more fighters to be embarked on the carriers.

Whilst no new battleships were laid down until the *King George V* class, the Royal Navy had to rely on the *Nelson*, *Rodney* and rebuilt ships as the core of its battle fleet.[28] In 1937 *Warspite* emerged from a four-year refit virtually as a new ship, with new engines and boilers, a completely new bridge, and improved armour and armament which included a completely new anti-aircraft battery and modifications to its 15-inch guns to give them a maximum range of 32,000 yards. In the same year, *Queen Elizabeth* and *Valiant* were taken in hand for similar modifications, although their 6-inch batteries were to be replaced with 20 4.5-inch dual-purpose guns in countersunk turrets. The battle cruiser *Renown* had been virtually rebuilt, although the outbreak of war was to stop similar rebuilds of the *Hood* and *Repulse*. Similarly, war and the shortage of 5.25-inch dual-purpose guns also stopped plans to replace the *Nelson*'s and *Rodney*'s mixed 4.7-inch and 6-inch secondary armament.

Regarding cruisers, the legacy of the London Naval Treaty was, as First Sea Lord Madden insisted, modern, light cruisers of the *Leander* and *Arethusa* classes, the former of 7,250 tons and armed with eight 6-inch guns, and the latter 5,220 tons and armed with six 6-inch guns. These ships fitted the original requirement for modern ships able to operate in the Far East, but they were still undergunned compared to foreign designs. So when, in 1935, the Admiralty came to decide on the next generation of cruisers, it settled on a design which reflected the preference to match the Japanese 8-inch cruisers with rapid-firing 6-inch guns. These were the *Town* class, with a displacement of 9,700 tons and an armament of 12 6-inch guns, followed by two improved *Edinburgh* class and the smaller *Colony* class which carried the same armament on a displacement of 8,000 tons. Another light cruiser class, the *Dido* class, was armed with 5.25-inch dual-purpose guns to augment the modified *C-class* anti-aircraft cruisers' ability to provide a heavy anti-aircraft barrage.

Destroyer building was similar. London had imposed a limit of 1,500 tons displacement and the Admiralty had opted for the maximum number of ships, building the almost identical *A* to *I* classes, each armed with four 4.7-inch guns and eight to ten 21-inch torpedo tubes. It became apparent that these ships were going to be at a disadvantage in a gun duel with foreign ships, and so from 1934 the *Tribal* class was introduced, which carried a heavy armament of eight 4.7-inch guns and only five 21-inch torpedo tubes. The newer destroyers built in the run up to the war were the *J* to *N* classes, smaller than the *Tribals* but with six 4.7-inch guns in double mounts, to provide a heavier broadside, and ten 21-inch torpedo tubes, to give a large torpedo spread. There were also the small escort destroyers of the *Hunt* class, built primarily for escort work.

Germany's renunciation of the Versailles Treaty in March 1935 had led to a change of focus, shifting naval strategy away from the China Sea and back to the North Sea again. Concerns over Britain's ability to deploy large fleets simultaneously in the Far East and home waters led to the 1935 Anglo-German Naval Treaty, an attempt by the government to limit German naval expansion until the Royal Navy's new construction plans were completed, although it was recognized that by 1940 the Royal Navy would still need significant naval forces in home and Mediterranean waters to match both the German and Italian navies, not to mention a French ally and luck.

As the Japanese became more aggressive in China and Germany increasingly appeared a threat to European stability, so the British naval position continued to deteriorate and, by 1937, the Naval Staff was pointing out that the Far Eastern strategy was now definitely a hostage to fortune:

> Recent indications have shown clearly that there is doubt whether under existing political conditions in Europe and with the rise of the German navy, we should, in fact, be able to send an adequate fleet to the Far East if a menace were to arise in that area.[29]

By the time of the Munich Crisis in 1938, the Committee of Imperial Defence felt that, even with a rearmament programme in place, the Royal Navy would be hard put to retain a numerical superiority over the combined German, Italian and Japanese fleets, although this did not stop the First Sea Lord, Sir Roger Backhouse, from reassuring the Australian High Commissioner that in the event of a war with Japan, seven battleships would be sent out to Singapore, despite his lack of faith in the strategy.[30]

Backhouse believed that Japan would only become involved in a war with Britain when Britain itself was embroiled in a European war, which would make the despatch of a battle fleet difficult. Sending even a small fleet for an unspecified time would be dangerous. European waters would be left without a naval presence whilst the seven ships in the Far East would offer no deterrent, only a tempting target for the Japanese.

Instead, Backhouse sought to realign the Admiralty's war plans to a more realistic set of objectives to deal with the German, Italian and Japanese threats. Drax revised the British war plans, on traditional lines, placing more emphasis on fighting to defend European and Mediterranean waters, not those of the Far East, whose safety was now to be guaranteed by a strong fleet in home waters, and not at Singapore. Instead, if possible, a small squadron, not unlike the pre-1914 'fleet unit' or the 1923 'Peace Fleet', of three battle cruisers was envisaged for disrupting Japanese fleet and troop convoy movements, until the arrival of a battle fleet. Backhouse's letter to Admiral Sir Ragnar Colvin, the First Naval Member for Australia, in 1939 tacitly acknowledged that with no ships to spare to send eastwards, the Singapore strategy had been, at best, radically recast, and at worst, abandoned. Only in the event of Japan entering or declaring war on Britain would it be in its national interests to send a fleet eastwards, but the strength and composition of that fleet would have to be determined by other operational requirements. Perhaps not explicitly, the Royal Navy was increasingly looking towards the United States Pacific Fleet as the provider of the battleship force that was needed at Singapore. Effectively though, the Far Eastern naval strategy as envisaged in War Memorandum (Eastern) faded in importance between 1931 and 1939, and even had the Japanese blockaded Tientsin, in June 1939, before Britain was embroiled in a world war, it is unlikely that an adequate fleet could have been sent eastwards. Conditions in Europe made it too much of a gamble. Only in 1941, when Churchill and Pound disagreed over the composition of the future detachment for Singapore, was the Royal Navy actually able to spare some vessels, and if the still neutral Americans had agreed to play a greater role covering the Atlantic convoys, or to send their battle fleet to Singapore, the strategy envisaged in War Memorandum (Eastern) could have been implemented more fully. As it was, the force sent was a hostage to fortune, its effectiveness largely negated by the Japanese incursions into French Indo-China and Malaya, and it was speedily sunk by Japanese naval air power within days of the outbreak of war.

Nevertheless, the tactical foundations laid by the Royal Navy in the 1920s were evident once war was declared in 1939. In the battles the Royal Navy fought, such as the Battle of the River Plate, the tactical ideas and developments, such as dispersing units, concentrating gunnery and aggressive handling of ships, were all evident. The River Plate action has been described by Eric Grove as a reflection of the inter-war flexible approach to command, and the first victory of the 'Beatty school of thought', building both on the Royal Navy's 'tradition of excellence' and the inter-war training. It was just the sort of action the navy had trained for, a surface action against an enemy raider at ranges of about 15 miles. Harewood's four cruisers were a powerful force, pitted against what was essentially a failed attempt to

CONCLUSION

design a cruiser with very heavy guns, at the expense of armour, and the rate of fire from the *Ajax* and *Achilles*, utilizing 'paired firing', caused a lot of splinter damage, especially to the *Graf Spee*'s rangefinder. The action showed that a supposedly inferior force could attack and win, and that the Royal Navy was now consciously aggressive.[31]

The glaring omission, when comparing the training and doctrines of 1919–31 with the realities of the war, was anti-submarine warfare, which had not received a high priority since 1918. Plans had been made to produce convoy escorts, and in 1939 and 1940 the first of the *Flower*-class corvettes were being built. But no one in 1939 could have foreseen the fall of France, the sudden availability of Atlantic bases for the *U*-Boats, or the development of submarine tactics which changed the whole nature of the war at sea.

The fall of France in 1940 also drastically altered the naval balance, increasing the demands on the Royal Navy's scarce resources, whilst Japan's occupation of the airfields in Indo-China brought their aircraft into striking range of Singapore. By the time of the Japanese declaration of war in 1941, naval losses up to that point – two battleships, one battle cruiser, three aircraft carriers and numerous cruisers and destroyers – meant that there was little hope of sending the fleet to the Far East without, perhaps, abandoning the Mediterranean Sea, a strategically sound but politically unacceptable action to many.[32] Faced with abandoning either the Mediterranean, or the Far East, the government on Churchill's insistence deployed a deterrent force and concentrated the China Squadron on Singapore.

If by 1931, and certainly by 1941, the ships necessary to send a battle fleet to the Far East were unavailable, why then were the Singapore strategy and War Memorandum (Eastern) important for the Royal Navy? The answer for most of the 1920s lay in the need to be able to justify a large battle fleet. Strategy was not an exclusively wartime concern and the need for strategy did not disappear in 1918,

> because military forces continue to exist and strategy controls their maintenance, future and training. Strategy must concern itself with possible future employment and can never be completely extinguished ... But strategy, though it continues to exist in uncertain and vacillating form in peacetime will not always find in the government's action the support, nourishment and inspiration that it has the right to expect.[33]

In other words, successive governments had to be persuaded that there was a real threat to imperial security which could only be answered by the provision of a large battle fleet and a naval base at

Singapore. Given that a navy took years to develop, that war with Japan was a possibility meant the Admiralty needed to have a battle fleet on hand. Long-term naval procurement and planning could not take place without a strategic justification, and by identifying a Japanese threat and laying down strategical plans, the Admiralty could also start planning for a navy to carry these plans out. Developing a naval base at Singapore and maintaining and supplying a fleet there had always been acknowledged as extremely difficult, right from the outset, so was it the case that having a large battle fleet, necessary to fight the Japanese, outweighed the practicalities of getting it to Singapore?[34] Perhaps here lies the true value of War Memorandum (Eastern): a dialogue between the strategists and those planning the fleet of the future. Having an enemy and a war plan would undoubtedly help to justify the need for higher levels of expenditure on the navy than perhaps the government would otherwise have contemplated.

Whatever the realities of a Japanese threat to British possessions in the Far East, the Japanese, as the only likely enemy, provided the strategical focus as well as the justification for a large battle fleet. If such a war broke out, then the fleet would need to be handled aggressively in battle and, once trained to fight the Japanese, the fleet could fight any naval opponent; fighting a battle fleet action against the Japanese would be little different from fighting against the Italians. Capital ships, cruisers, aircraft, destroyers and submarines were all given tactical roles to play in a fleet battle which emphasized victory, maximizing the tactical roles, encouraging subordinate commanders to use their initiative and to make their mistakes when it was not crucial, in peacetime. Exercises were a way of trying to bring this about. In moving away from the rigidity of Jellicoe's Grand Fleet Battle Orders, the Royal Navy was rediscovering a philosophy that had as its aim the disruption of the enemy's cohesion, taking the war to them, and forcing them to respond, creating a situation where constant maritime pressure would bring about the collapse of Japanese morale.[35]

The strategy, War Memorandum (Eastern), provided the justification for a large battle fleet equipped with the tactical skills and expertise to defeat an enemy, which served the Royal Navy well during the Second World War. In many ways the strategy itself was questionable, depending on too many variables: a fully functioning, impregnable base at Singapore, the availability of ships and supplies, the compliance of the other European naval powers and, of course, the Japanese.

This much should have been obvious to all sides. It may have been to the Japanese and it certainly was to G. Lambert, the MP for South

CONCLUSION

Moulton, who indicated in a House of Commons debate, on 19 July 1923, that as the Grand Fleet had been unable to go 300 miles to the German coast in the First World War, it was unlikely to be able to go 10,000 miles to the Far East in the next.[36] The inference here is that obviously it was impractical, and the lessons of the Great War should have suggested this very clearly. But, ultimately, this was perhaps not where the main importance of War Memorandum (Eastern) lay. The Royal Navy made significant improvements between the wars, and maintained their position as the world's largest and, by inference, most powerful navy. Ships were constructed or refitted with more armour protection, gunnery had improved with the introduction of the AFCT, aircraft promised to improve spotting and reconnaissance, and to make a contribution to battle; it was publicly claimed that the submarine had been neutralized by ASDIC, although, in reality, ASDIC was still a very limited and imperfect means of detecting submarines, a fact that had been shown in exercises and was acknowledged inside the Service.

The Battle Instructions advocated a much more fluid, aggressive approach to battle, and exercises seemed to emphasize the greater use of initiative. The Royal Navy did not doubt its ability to fight and win a naval battle against the Imperial Japanese Navy in the years between 1919 and 1931, and it was only the changing European and Mediterranean situation, and the knowledge that the Royal Navy would be overstretched that caused the planning to change. The development of a naval base at Singapore remained a key part of this overall strategy, as there had to be a base capable of providing for the needs of a battle fleet, whether or not this flew the White Ensign.

There were several flaws in the British Far Eastern strategy, however. Sending a battle fleet eastwards was entirely dependent on a peaceful Europe. Logistically, it would have been a massive undertaking to get a battle fleet to Singapore and once there the fleet would be committed to remaining in eastern waters for an indefinite time, until Japan was defeated. During this time, possibly three to four years, the European situation would need to remain stable. Richmond, in his critiques of War Memorandum (Eastern), pointed out the small superiority that the Royal Navy would have had in gun power, and the risks the country would be running deploying the fleet to Singapore. But both he and Drax were reasonably confident that with superior training the Royal Navy would emerge victorious. Despite the Royal Navy's feeling increasingly unable even to match Japanese naval forces during the 1930s, it did still feel that Japanese ships might possibly have been, given the Japanese obsession with secrecy, inferior to the British ships, whilst racially, the British placed the Japanese on a par with the Italians, believing that they would fight, but would soon give way.

But the only true test of a strategy and of tactics can be whether they work in wartime and, in this respect, War Memorandum (Eastern) remained theoretical. Certainly it was an ambitious plan, probably unrealistic, but one that provided the context for the Royal Navy to develop both the future fleet levels and training throughout the 1920s. Political developments in Europe changed the strategic context in the 1930s, and by 1942 and the fall of Singapore the context had changed again, as had the strategy, to reflect the fact that Britain was fighting a world war against three enemies. Given all of these factors, it is reasonable to question the purpose of War Memorandum (Eastern). The ultimate criterion for judging success, a war as envisaged by the plan, never happened. Even if it had, it was unlikely that the Royal Navy would have had all of the fuel and supply requirements in place to implement it and responses to situations would have had to be improvised. Despite this, the war planning was successful in one important way. It did provide a long-term planning focus for the Royal Navy; and because of its very difficulty, it provided the focus that the Admiralty needed to determine the composition of the future battle fleet, the means of fighting, the levels of logistical support that would be needed, a viable, realistic enemy, and a focus for planning and political debate.

NOTES

1. E. Grove and P. Hore, *The Fundamentals of British Maritime Doctrine, BR 1806* (HMSO, London, 1995), p. 77.
2. Rear Admiral R. Menon, *Maritime Strategy and Continental Wars* (Frank Cass, London, 1998), pp. 22–3.
3. I. Gow, 'Anglo-Japanese Naval Relations Prior to 1931', in I. Nish (ed.), *Anglo-Japanese Naval Relations* (London School of Economics, London, 1985), pp. 20–1.
4. PRO, ADM 116/3125, 'War Memorandum (Eastern), August 1924', p. 12.
5. C. Bell, 'The Royal Navy, War Planning and Intelligence', in Siegel and Jackson (eds), *Intelligence and International Systems 1870–1970*.
6. Louis, *British Strategy in the Far East, 1919–1939*, p. 142.
7. C. Bell, *The Royal Navy, Sea Power and Strategy between the Wars* (Macmillan, London, 2000), p. 97.
8. J.R. Ferris, 'The Decade of British Maritime Supremacy 1919–1929', in Neilson and Kennedy (eds), *Far Flung Lines*, p. 137.
9. Ibid.
10. M. Murfett 'Living in the Past: A Critical Re-examination of the Singapore Naval Strategy, 1918–1941', *War in Society*, Vol. 11, No. 1, May 1993, p. 81.
11. Ibid., p. 82.
12. Bell, 'The Royal Navy, War Planning and Intelligence', p. 27.

CONCLUSION

13. Ferris, 'The Last Decade', p. 140.
14. Gow, 'Anglo-Japanese Naval Relations', p. 24.
15. Ibid.
16. Murfett, 'Living in the Past', p. 81; NMM, Richmond Papers, RIC 11/2, 'The Battleship', 14 February 1923.
17. PRO, ADM 116/3125, p. 12.
18. Gordon, 'The Rules of the Game' and 'The Doctrine Pendulum', in *The Hudson Papers*, Vol. I.
19. Hughes, *Fleet Tactics*, pp. 35–7.
20. Captain C. Pehl, USN, 'Through a Glass, Darkly', *United States Naval Institute Proceedings*, Vol. 122, September 1996, p. 57.
21. G. Till, 'Retrenchment, Rethinking, Revived, 1919–1937', in Hill (ed.), *Oxford Illustrated History of the Royal Navy*, p. 343.
22. Moretz, *The Royal Navy*, p. 239.
23. David Hobbs, Curator, Fleet Air Arm Museum, Yeovilton, in telephone conversation with author, 25 May 2001, and at Portsmouth Conference, 'Classic Naval Battles Revisited: Taranto', Naval Academy, Portsmouth, 16 February 2002.
24. J. Sumida, 'Technology and Naval Combat', in O'Brien (ed.), *Technology and Naval Combat*, p. 138; C.H. Bailey (ed.), *The Life and Times of Admiral Sir Frank Twiss: Social Change in the Royal Navy 1924–1970* (Royal Naval Museum/Sutton Publishing, Stroud, 1996), p. 47.
25. Sumida, 'Technology and Naval Combat', pp. 139–40.
26. G. Till, 'Air Power and the Battleship', in Ranft, *Technical Change and British Naval Policy*, p. 116; Friedman, *British Carrier Aviation*, p. 160.
27. J.B. Hattendorf, R.J.B. Knight, A.W.H. Pearsall, N.A.M. Rodger and G. Till (eds), *British Naval Documents, 1204–1960* (Navy Records Society/Scolar Press, Aldershot, 1993), 'The Building of the *Ark Royal*: Sir Bolton Eyres-Monsell, First Lord of the Admiralty to Sir Philip Sassoon, Under Secretary of the Air Ministry, 31 January 1934', pp. 944–5.
28. Chesneau (ed.), *Conway's All The World's Fighting Ships, 1922–1946* (1980); various copies of the annual, *Jane's Fighting Ships* (Sampson Low, or David and Charles reprints) provide details.
29. Murfett, 'Living in the Past', p. 87.
30. Ibid., p. 90.
31. E. Grove, 'Classic Naval Battles Revisited: The Battle of the River Plate', Portsmouth Conference, The Naval Academy, Portsmouth, 16 February 2002.
32. PRO, CAB 16/209, 'CID 6th Meeting, 17 April 1939'; ADM 116/3900, 'Strategical Aspects of the Situation in the Mediterranean'; Captain C. Page (ed.), *The Royal Navy and the Mediterranean: Volume 1, September 1939–October 1940* (Whitehall History Publishing/Frank Cass, London, 2002), p. 3; Brodhurst, *Churchill's Anchor*, pp. 107–9, all discuss the strategic situation with regard to the Mediterranean and Far East in the event of a war with Italy.
33. Captain R. Castex, *Strategic Theories*, ed. E.S. Keisling (Naval Institute Press, Annapolis, MD, 1994), p. 250.
34. NMM, Richmond Papers, RIC11/1, 'Supply of Fleets and Bases', Captain N.F. Lawrence, 22 March 1922.
35. Gordon, 'Rules of the Game', pp. 562–601.
36. Murfett, 'Living in the Past', p. 97.

Select Bibliography

MANUSCRIPT AND OTHER SOURCES

Churchill College Archive, Churchill College, Cambridge

Admiral Sir Frederick Charles Dreyer Papers.
Captain The Honourable R.A.R. Plunkett-Ernle-Erle Drax DSO Papers.
Admiral J.H. Godfrey RN Papers.
Admiral of the Fleet Sir Dudley Pickman Rogers Pound RN Papers.
Captain Stephen Roskill RN Papers.

Imperial War Museum, London

Captain R.C. Bayne RN Papers.
Rear Admiral Ross Papers.

National Maritime Museum (NMM), Greenwich

Admiral Sir Herbert Richmond Papers.

Naval Historical Branch, Ministry of Defence, London

Unpublished reports from the Naval Attaché, Tokyo, to the Admiralty:
 'Naval Notes 1921'.
 'Naval Notes 1929'.
 'Naval Notes 1930'.
'Staff Appreciation of the War in the Pacific', Vol. I.
'Historical Précis on the Japanese Navy and Japanese Far Eastern Policy', C.G. Stuart, 1924.
CB 3016/30 'Progress in Tactics', 1930.

CB 3016/31 'Progress in Tactics', 1931.
CB 3016/37 'Progress in Tactics', 1937.
The Godfrey Papers: Staff College Lectures 1924–25, Royal Navy Staff College, Greenwich.
'Outline of Naval Armaments and Preparations for War, Part I.' Japanese Monograph 145 (translation), HQ, Far Eastern Command, Military History Section, Special Staff Japanese Research Division, 1945.

Public Records Office (PRO), Kew, London

ADM 1/8570/287 'British Imperial Naval Bases in the Pacific' April 1919.
ADM 1/8571/295 'The Naval Situation in the Far East' October 1919.
ADM 1/8948 'The Naval Situation of the British Empire in the event of war between Japan and the US of America'.
ADM 116/1773 'Memorandum to the War Cabinet 1919'.
ADM 116/2060 'Post War Questions Committee'.
ADM 116/2247 'Empire Naval Policy – Imperial Conference 1925'.
ADM 116/2335 'Organization of Mobile Naval Bases, 30 August 1920'.
ADM 116/2480 'Mediterranean Fleet Spring and Summer Cruise 1926'.
ADM 116/2684 'Manual of Mobile Naval Base Defence Organization', 1927–29, 1931–34.
ADM 116/2746 'Basis of British Naval Strategy; Memorandum by Admiral of the Fleet Sir Charles Madden, 17th January 1930'.
ADM 116/2764 'London Naval Conference for the Reduction of Armaments, 1930, Secret Memoranda Prepared for Use of the British Delegation Only'.
ADM 116/3118 'War Memorandum (Eastern) 1931–32'.
ADM 116/3121 'Remarks on the Report of the Singapore Conference, 1925'.
ADM 116/3125 'War Memorandum (Eastern) August 1924'.
ADM 116/3134 'Mediterranean Fleet Preparations and Readiness for War, 6 January 1927'.
ADM 116/3167 'Pacific War Plans and Naval Problems'.
ADM 116/3195 'Redistribution of the Fleet to Meet Changes in the World Political Situation and a Possible War Threat in the Middle and Far East, 1923'.
ADM 116/3438 '1923 Imperial Conference; Admiralty Policy with regard to Dominion Navies'.
ADM 116/3445 'The Washington Conference on Limitation of Armaments and Far Eastern Policy 1921–22, Vol. 1'.
ADM 116/3673 'War Memorandum (Eastern) 1938'.

ADM 116/4444 'Passage of the Fleet to the Far East: Secret Fuelling Instructions'.
ADM 186/53 'Cruiser Manual, 1929'.
ADM 186/66 'Naval War Manual, October 1925'.
ADM 186/72 'Battle Instructions, 1922–1927'.
ADM 186/75 'Fighting Instructions, 1928'.
ADM 186/78 'CB3011 War Games Rules, 1929'.
ADM 186/143 'Selected Reports of Exercises, Operations and Torpedo Practices in HM Fleet, Summer and Autumn 1927'.
ADM 186/144 'Selected Reports of Exercises, Operations and Torpedo Practices in HM Fleet (Vol. 2) and Annual Summary of Tactical Progress During 1928'.
ADM 186/145 'Exercise and Operations 1929, Vol. 1'.
ADM 186/146 'Exercises and Operations 1929, Vol. 2'.
ADM 203/47 'The Possibility of Losing Hong Kong in a War with Japan in 1926. Lieutenant Commander R.R. Stewart, RN'.
ADM 203/86 'The Relief of Alboran Island, March 1928'.
ADM 203/84 'Combined Naval and Military Exercise Carried Out on Salsette Island, Bombay, December 1924'.
ADM 203/90 'Strategical Exercise MZ – Atlantic and Mediterranean Fleets, March 1929'.
ADM 205/3 'First Sea Lord's Records 1939–1945'.
Air 9/2 'Naval Cooperation and Fleet Air Arm 1914–1939'.
CAB 16/20/9 'Committee of Imperial Defence Meetings 1939'.
CAB 16/37/1 'Sub Committee on the Question of the Capital Ship in the Navy, December 1920/January 1921, Vol. I'.
CAB 16/37/2 'Sub Committee on the Question of the Capital Ship in the Navy, December 1920/January 1921, Vol. II'.
FO371/5365 'Naval Attaché's Visit to Imperial Dockyards and Private Shipping Establishments in Japan, 12 July 1920'.

Royal Naval Museum, Portsmouth

Royal Naval Museum Manuscript 1994/332; Lecture Précis Course Notes, 'Senior Warrant Officers' Course, Royal Naval College, Greenwich, Autumn 1925'.

US Naval War College, Newport, Rhode Island, United States

Laning, Captain H., USN: *'The Battle of Emerald Bank* as manoeuvred at the US Naval War College by the Class of 1923. History and Tactical Critique.' US Naval War College RG 4.

BIBLIOGRAPHY

'*The Battle of Sable Island* as manoeuvred at the US Naval War College by the Class of 1924. History and Tactical Critique.' US Naval War College RG 4.

PRINTED ORIGINAL SOURCES

Ashton, Major General Sir G. *Combined Operations* (*Journal of the Royal United Services Institution*, 1919, reprinted Pallas Armata, 1997).

D'Eyncourt, Sir Eustace Tennyson, *Naval Construction During the War* (*Transactions of the Institute of Naval Architects*, 9 April 1919, reprinted Pallas Armata, 2000).

D'Eyncourt, Sir Eustace and Fea, Major L., Royal Italian Navy, *Consequences of the Washington Conference* (*Transactions of the Institute of Naval Architects*, 4 July 1922, reprinted Pallas Armata, 2000).

Faure, Lieutenant C.M., RN, *The Influence of the Submarine in Naval Warfare in the Future* (*Journal of the Royal United Services Institution*, Vol. LXIV, No. 456, 1919, reprinted Pallas Armata, 1997).

Fea, Major L., Royal Italian Navy, *Some of the Consequences of the Washington Conference* (*Transactions of the Institute of Naval Architects*, 4 July 1922, reprinted Pallas Armata, 2000).

Genda, M. 'Evolution of Aircraft Carrier Tactics of the Imperial Japanese Navy', unpublished monograph.

Halpern, P. (ed.), *The Keyes Papers, Vol. II, 1919–1938* (Navy Records Society, London, 1980).

Hannay, D. *Sea Power in the Pacific* (*The Edinburgh Review*, 1921, reprinted, Pallas Armata, 1997).

Hattendorf, J.B., Knight, R.J.B., Pearsall, A.W.H., Rodger, N.A.M. and Till, G. (eds), *British Naval Documents 1204–1960* (Navy Records Society, London, 1993).

Howarth, S. and Law, D. *The Battle of the Atlantic 1939–45: The 50th Anniversary International Conference* (Greenhill, London, 1994).

King Hall, Lieutenant W.S. *The Influence of the Submarine in Naval Warfare in the Future*, pamphlet (*Journal of the Royal United Services Institution*, Vol. LXIV, No. 445, 1919, reprinted Pallas Armata, 1997).

Norman, Captain A.H., RN, *The Advantages and Disadvantages of a Separate Air Force for the Royal Navy* (*Journal of the Royal United Services Institution*, 1923, reprinted Pallas Armata, 1997).

Ranft, B. (ed), *The Beatty Papers, Vol. I, 1902–1918 and Vol. II, 1916–1927* (Navy Records Society, London, 1993).

Richard, Capitaine de Corvette, *Jutland and the Principles of War* (*Journal of the Royal United Services Institute*, 1921, reprinted Pallas Armata, 1997).

Roskill, Captain S.W. (ed.), *The Naval Air Service 1908–1918* (Navy Records Society, London, 1969).

Simpson, M. (ed), *Anglo-American Naval Relations 1917–1919* (Navy Records Society, London, 1991).

—— *The Somerville Papers: Selections from the Private and Official Correspondence of Admiral of the Fleet Sir James Somerville, GCB, GBE, DSO* (Navy Records Society/Scolar Press, Aldershot, 1995).

Sumida, J.T. *The Pollen Papers* (Navy Records Society, London, 1984).

Temple Patterson, A. (ed.), *The Jellicoe Papers, Vol. I, 1893–1916* and *Vol. II, 1916–1935* (Navy Records Society, London, 1968).

Tracy, N. *The Collective Naval Defence of the Empire, 1900–1940* (Navy Records Society, Ashgate, Aldershot, 1997).

SECONDARY SOURCES

Archibald, E.H.H. *The Fighting Ship in the Royal Navy 1897–1984* (Blandford, London, 1987).

Baer, G.W. *One Hundred Years of Sea Power: The US Navy 1890–1990* (Stanford University Press, Stanford, CA, 1994).

Bailey, C.H. (ed.), *Social Change in the Royal Navy, 1924–1970: The Life and Times of Admiral Sir Frank Twiss* (Alan Sutton, Stroud, 1996).

Ballantyne, I. *Warspite* (Pen and Sword/Leo Cooper, London, 2001).

Ballard, Vice Admiral G.A., RN, *The Influence of the Sea on the Political History of Japan* (John Murray, London, 1921).

Barnett, C. *Engage the Enemy More Closely: The Royal Navy in the Second World War* (Norton, New York, 1991).

Beesly, P. *Very Special Admiral: The Life of Admiral J.H. Godfrey, CB* (London, 1980).

Bell, C.M. *The Royal Navy, Seapower and Strategy Between the Wars* (Macmillan, London, 2000).

Bell, P.M.H. *The Origins of the Second World War in Europe* (Longman, London, 1986).

Bond, B. *British Military Policy Between the Two World Wars* (Clarendon Press, Oxford, 1980).

Brice, M. *The Royal Navy and the Sino-Japanese Incident, 1937–1941* (Ian Allan, Shepperton, 1973).

Brodhurst, R. *Churchill's Anchor: The Biography of Admiral of the Fleet Sir Dudley Pound, OM, GCB, GCVO* (Pen & Sword, London, 2000).
Burt, R.A. *British Battleships 1919-1939* (Arms & Armour Press, London, 1993).
Burton, A. *The Rise and Fall of British Shipbuilding* (Constable, London, 1994).
Bywater, H. *A Searchlight on the Navy* (Constable, London, 1934).
Cain, P.J. and Hopkins, A.G. *British Imperialism: Crisis and Deconstruction, 1914-1990* (Longman, London, 1993).
Campbell, J. *Jutland: An Analysis of the Fighting* (Conway Maritime Press, London, 1986).
Castex, R. *Strategic Theories*, ed. E.S. Kiesling (Naval Institute Press, Annapolis, MD, 1994).
Chalmers, W.S. *The Life and Letters of David Beatty, Admiral of the Fleet* (Hodder & Stoughton, London, 1951).
Chatfield, Admiral of the Fleet Lord E. *The Navy and Defence: Vol. I of the Autobiography of Admiral of the Fleet Lord Chatfield* (Heineman, London, 1942).
—— *It Might Happen Again: Vol. II of the Biography of Admiral of the Fleet Lord Chatfield* (Heineman, London, 1947).
Chesneau, R. (ed.), *Conway's All the Worlds' Fighting Ships, 1922-1946* (Conway Maritime Press, London, 1986).
Chit Chung, O. *Operation Matador: Britain's War Plans Against the Japanese, 1918-1941* (Times Academic Press, Singapore, 1997).
Compton Hall, R. *Submarines and the War at Sea 1914-18* (Macmillan, London, 1991).
Corbett, J. *Some Principles of Maritime Strategy*, ed. E. Grove (Brassey's, London, 1988).
Cowman, I. *Dominion or Decline: Anglo-American Naval Relations in the Pacific, 1937-1941* (Berg, Oxford, 1996).
Cunningham of Hyndhope, Admiral of the Fleet Viscount. *A Sailor's Odyssey: The Autobiography of Admiral of the Fleet Viscount Cunningham of Hyndhope* (Hutchinson, London, 1951).
Dingman, R. *Power in the Pacific: The Origins of Naval Arms Limitation, 1914-1922* (University of Chicago Press, Chicago, IL, 1976).
Edmonds, M. (ed.), *100 Years of 'The Trade'* (CDISS, University of Lancaster, 2001).
Elphick, P. *Singapore: The Pregnable Fortress* (Hodder & Stoughton, London, 1995).
Evans, D.C. and Peattie, M.R. *Kaigun: Strategy, Tactics and Technology in the Imperial Japanese Navy 1887-1941* (Naval Institute Press, Annapolis, MD, 1997).

Ferris, J.R. *The Evolution of British Strategic Policy 1919–1926* (Macmillan, London, 1989).

Fiske, Rear Admiral Bradley A., USN. *The Navy as a Fighting Machine*, ed. W. Hughes (Naval Institute Press, Annapolis, MD, 1988).

Forester, C.S. *The Ship* (Penguin, London, 1971).

Fraser, D. *And We Shall Shock Them: The British Army in the Second World War* (Hodder & Stoughton, London, 1983).

Friedman, G. and Lebard, M. *The Coming War with Japan* (St Martins Press, New York, 1991).

Friedman, N. *British Carrier Aviation: The Evolution of the Ships and their Aircraft* (Conway Maritime Press, London 1988).

Gardiner, R. and Brown, D. (eds), *The Eclipse of the Big Gun – The Warship 1906–1945* (Conway Maritime Press, London, 1992).

Gill, G.H. *Royal Australian Navy, Vol. I, 1939–1942* (Collins, Sydney, 1957).

Gilmour, D. *Curzon* (Macmillan, London, 1995).

Goldman, E.O. *Sunken Treaties: Naval Arms Control Between the Wars* (Penn State Press, University Park, PA, 1994).

Goldrick, J. *The King's Ships Were at Sea: The War in the North Sea, August 1914–February 1915* (Naval Institute Press, Annapolis, MD, 1984).

Goldstein, E. and Maurer, J. (eds), *The Washington Conference 1921–22: Naval Rivalry, East Asian Stability and the Road to Pearl Harbor* (Frank Cass, London, 1994).

Gordon, G.A.H. *British Seapower and Procurement Between the Wars: A Reappraisal of Rearmament* (Naval Institute Press, Annapolis, MD, 1988).

——*The Rules of the Game: Jutland and British Naval Command* (John Murray, London, 1996).

Gray, C. and Barnett, R.W. (eds), *Seapower and Strategy* (Tri-Service Press, London, 1989).

Gray, R. (ed.), *Conway's All The World's Fighting Ships 1906–1921* (Conway Maritime Press, London, 1985).

Grove, E. *Fleet to Fleet Encounters* (Arms & Armour Press, London, 1991).

Grove, E. (ed.), *The Battle and the Breeze: The Naval Reminiscences of Admiral of the Fleet Sir Edward Ashmore* (Alan Sutton, Stroud, 1997).

Grove, E. and Hore, P. *The Fundamentals of British Maritime Doctrine, BR 1806* (HMSO, London, 1995).

Haggie, P. *Britannia at Bay* (Oxford University Press, Oxford, 1981).

Hall, C. *Britain, America and Arms Control, 1921–37* (Macmillan, London, 1987).

BIBLIOGRAPHY

Halpern, P. *A Naval History of World War One* (UCL Press, London, 1994).

Hattendorf, J.B. (ed.), *Mahan on Naval Strategy* (Naval Institute Press, Annapolis, MD, 1991).

Higham, R. *Armed Forces in Peacetime: Britain 1815–1940, A Case Study* (G.T. Foulis, London, 1962).

Hill, J.R. (ed.), *The Oxford Illustrated History of the Royal Navy* (BCA, London, 1995).

Hodges, P. *The Big Gun* (Conway Maritime Press, London, 1989).

Honan, W. *Bywater, the Man Who Invented the Pacific War* (Macmillan, London, 1991).

Hough, R. *Former Naval Person* (Weidenfeld & Nicolson, London, 1987).

—— *The Great War at Sea 1914–1918* (Oxford University Press, Oxford, 1983).

Howarth, S. *Morning Glory: A History of the Imperial Japanese Navy* (Hamish Hamilton, London, 1983).

—— *To Shining Sea: A History of the United States Navy 1775–1991* (Weidenfeld & Nicolson, London, 1991).

Hughes, W. *Fleet Tactics: Theory and Practice*, (Naval Institute Press, Annapolis, MD, 1986).

Hunt, B. *Sailor–Scholar: Admiral Sir Herbert Richmond 1871–1946* (Wilfred Laurier University Press, Waterloo, Ontario, 1982).

Ireland, B. *Cruisers* (BCA, London, 1981).

Iriye, A. *The Origins of the Second World War in the Pacific* (Longman, London, 1987).

Jellicoe, Admiral of the Fleet Viscount. *The Grand Fleet 1914–1918: Its Creation, Development and Work* (Cassell, London, 1919).

Kaufman, R.G. *Arms Control during the Pre-Nuclear Age: The United States and Arms Limitation between the Two World Wars* (Columbia University Press, New York, 1990).

Keegan, J. *The Price of Admiralty: War at Sea from Man of War to Submarine* (Hutchinson, London, 1988).

Kennedy, P. *The War Plans of the Great Powers 1880–1914* (Allen & Unwin, London, 1985).

—— *The Realities behind Diplomacy: Background Influences on British External Policy 1865–1980* (Fontana, London, 1985).

—— *The Rise and Fall of British Naval Mastery* (Macmillan, London, 1986).

—— *The Rise and Fall of the Great Powers* (Random House, New York, 1987).

—— *Strategy and Diplomacy* (Fontana, London, 1989).

King, E. and Muir Whitehill, W. *Fleet Admiral King: A Naval Record* (Eyre & Spottiswoode, London, 1953).

Lamb, R. *The Drift to War 1922–1939* (Bloomsbury, London, 1991).
Lambert, N.A. *Sir John Fisher's Naval Revolution* (University of South Carolina Press, Columbia, SC, 1999).
Layman, R.D. *Naval Aviation in the First World War: Its Impact and Influence* (Conway Maritime Press, London, 1996).
Layman, R.D. and McLaughlin, S. *The Hybrid Warship* (Conway Maritime Press, London, 1991).
Louis, W.R. *British Strategy in the Far East 1919–1939* (Clarendon Press, Oxford, 1971).
McMurtrie, F. (ed.), *Jane's Fighting Ships 1931* (David & Charles reprint, London, 1973).
Mahan, A.T. *The Influence of Sea Power upon History 1660–1805*, ed. E. Grove (Bison/Hamlyn, London, 1980).
Maiolo, J.A. *The Royal Navy and Nazi Germany, 1933–39: A Study in Appeasement and the Origins of the Second World War* (Macmillan, London, 1998).
Marder, A.J. *Portrait of an Admiral: The Life and Papers of Sir Herbert Richmond* (Harvard University Press, Cambridge, MA, 1952).
—— *From the Dreadnought to Scapa Flow: The Royal Navy in the Fisher Era, 1904–1919, Vol. 1: The Road to War, 1904–1914* (Oxford University Press, Oxford, 1961).
—— *The Anatomy of British Sea Power: A History of British Naval Policy in the Pre-Dreadnought Era 1880–1905* (Frank Cass, London, 1972).
—— *From the Dardanelles to Oran: Studies of the Royal Navy in War and Peace* (Oxford University Press, Oxford, 1974).
—— *Old Friends, New Enemies: The Royal Navy and the Imperial Japanese Navy, Vol. I* (Oxford University Press, Oxford, 1981).
Menon, Rear Admiral R. *Maritime Strategy and Continental Wars* (Frank Cass, London, 1998).
Middlebrook, M. and Mahoney, P. *Battleship, the Loss of the 'Prince of Wales' and 'Repulse'* (Penguin, London, 1979).
Miller, E. *War Plan Orange: The US Strategy to Defeat Japan 1897–1945* (Naval Institute Press, Annapolis, MD, 1991).
Moretz, J. *The Royal Navy and the Capital Ship in the Interwar Period: An Operational Perspective* (Frank Cass, London, 2002).
Morris, J. *Fisher's Face* (Penguin, London, 1996).
Murfett, M. (ed.), *The First Sea Lords: From Fisher to Mountbatten* (Praeger, Westport, CT, 1995).
Neidpath, J. *The Singapore Naval Base* (Clarendon Press, Oxford, 1981).
Neilson, K. and Kennedy, G. *Far Flung Lines: Studies in Imperial Defence in Honour of Donald Mackenzie Schurman* (Frank Cass, London, 1997).

Nicholls, B. *Statesmen and Sailors: Australian Maritime Defence 1870–1920* (B. Nicholls, Balmain, New South Wales, 1995).

Nish, I. *Alliance in Decline – A Study in Anglo-Japanese Relations 1908–1923* (University of London, London, 1972).

O'Brien, P.P. *British and American Naval Power: Politics and Policy 1900–1936* (Praeger, Westport, CT, 1998).

—— (ed.), *Technology and Naval Combat in the Twentieth Century and Beyond* (Frank Cass, London, 2001).

Overy, R.J. *The Inter-War Crisis 1919–1939* (Longman, London, 1994).

—— *The Origins of the Second World War* (Longman, London, 1995).

Overy, R.J. and Wheatcroft, A. *The Road to War* (Macmillan/BBC, London, 1989).

Padfield, P. *The Battleship Era* (MBC, London, 1972).

—— *Guns at Sea* (Hugh Evelyn, London, 1972).

—— *War Beneath the Sea: Submarine Conflict 1939–1945* (John Murray, London, 1995).

Page, Captain C. (ed.), *The Royal Navy and the Mediterranean, Vol. I: September 1939–October 1940* (Whitehall History Publishing/Frank Cass, London, 2002).

Parker, R.A.C. *Chamberlain and Appeasement: British Foreign Policy and the Coming of the Second World War* (Macmillan, London, 1993).

Parkes, O. *British Battleships 1860–1950: A History of Design, Construction and Armament* (Leo Cooper, London, 1966).

Parkes, O. and Prendergast, M. (eds), *Jane's Fighting Ships 1919* (David & Charles, London, 1969).

Parkes, O. and McMurtrie, F. (eds), *Jane's Fighting Ships 1924* (David & Charles, London, 1973).

Peattie, M.R. *Sunburst: The Rise of Japanese Naval Air Power, 1909–1941* (Naval Institute Press, Annapolis, MD, 2001).

Perla, P. *The Art of Wargaming* (Naval Institute Press, Annapolis, MD, 1990).

Polmar, N. *Aircraft Carriers – A Graphic History of Carrier Aviation and its Influence on World Events* (Doubleday, New York, 1969).

Ponting, C. *Churchill* (Sinclair-Stevenson, London, 1994).

Porter, B. *The Lion's Share: A Short History of British Imperialism, 1850–1995* (Longman, London, 1996).

Preston, A. *Destroyers* (Hamlyn, London, 1977).

—— *Aircraft Carriers* (Hamlyn, London, 1979).

—— *Cruisers: An Illustrated History* (Arms & Armour Press, London, 1980).

—— *Battleships* (Bison/Hamlyn, London, 1981).

—— *Submarines* (Bison/Hamlyn, New York, 1982).
Pugh, M. *Lloyd George* (Longman, London, 1994).
Pugh, P. *The Cost of Seapower: The Influence of Money on Naval Affairs from 1815 to the Present Day* (Conway Maritime Press, London, 1986).
Ranft, B. (ed.), *Technical Change and British Naval Policy 1860–1939* (Hodder & Stoughton, London, 1977).
—— (ed.), *Ironclad to Trident: 100 Years of Defence Commentary: Brassey's 1886–1986* (Brassey's, London, 1986).
Reckner, J. *Teddy Roosevelt's Great White Fleet* (Naval Institute Press, Annapolis, MD, 1988).
Reynolds, C. *The Fast Carriers: The Forging of an Air Navy* (Naval Institute Press, Annapolis, MD, 1968).
Robinson, Cdr Charles, RN and Ross, H.M. (eds), *Brassey's Naval and Shipping Annual* (W. Clowes & Son, London, 1931–36).
Rodger, N.A.M. *The Admiralty* (Terence Dalton, Lavenham, 1979).
Roskill, Captain S.W. *Naval Policy Between the Wars, Vol. I: The Period of Anglo-American Antagonism 1919–1929* (Collins, London, 1968).
—— *Admiral of the Fleet Earl Beatty: The Last Naval Hero: An Intimate Biography* (Collins, London, 1980).
Schurman, D.M. *The Education of a Navy: The Development of British Naval Strategic Thought, 1867–1914* (Robert E. Krieger, Malabar, FL, 1984).
Semmel, B. *Liberalism and Naval Strategy: Ideology, Interest and Sea Power During the Pax Britannica* (Allen & Unwin, London, 1986).
Smith, M. *British Air Strategy Between the Wars* (Oxford University Press, Oxford, 1984).
Smith, P. (ed.), *Government and the Armed Forces in Britain 1856–1990* (Hambledon Press, London, 1996).
Stephen, M. *The Fighting Admirals* (Leo Cooper, London, 1991).
Stephen, M. and Grove, E. *Sea Battles in Close Up: World War 2, Vol. 1* (Naval Institute Press, Annapolis, MD, 1993).
Stevens, D. (ed.), *Maritime Power in the 20th Century: The Australian Experience* (St Leonards, New South Wales, 1998).
—— *In Search of a Maritime Strategy: The Maritime Element in Australian Defence Planning since 1901* (Canberra Papers on Strategy and Defence No. 119, Strategic and Defence Studies Centre, Canberra, 1997).
Sturtivant, R. *British Naval Aviation: The Fleet Air Arm 1917–1990* (Naval Institute Press, Annapolis, MD, 1990).
Sumida, J.T. *In Defence of Naval Supremacy: Finance, Technology and British Naval Policy 1889–1914* (Routledge, London, 1993).

Tarrant, V.E. *Jutland: The German Perspective* (Arms & Armour Press, London, 1995).
Temple Patterson, A. *Tyrwhitt of the Harwich Force* (Macdonald, London, 1973).
Terraine, J. *Business in Great Waters: The U Boat War 1916–1945* (Leo Cooper, London, 1989).
Thorne, C. *Allies of a Kind: The United States, Britain and the War Against Japan* (Hamish Hamilton, London, 1978).
Thursfield, Rear Admiral H.G., RN, *Brassey's Naval and Shipping Annual* (W. Clowes & Son, London, 1937–1938).
Till, G. *Air Power and the Royal Navy* (Janes Publishing, London, 1979).
—— *Seapower; Theory and Practice* (Frank Cass, London, 1994).
Wells, Captain John. *The Royal Navy: An Illustrated Social History, 1870–1982* (Alan Sutton, Stroude, 1999).
White, C. *Victoria's Navy: The Heyday of Steam* (Kenneth Mason, London, 1983).

CONFERENCE PAPERS

University of Exeter

Baugh, D. 'Confounded by Perplexities: The Navy and British Defence Planning between the Wars', at conference on 'The Parameters of Naval Power in the Twentieth Century', 4–7 June 1994.

Society for Nautical Research/Centre for Security Studies, University of Hull

'The Battle of Jutland Reconsidered', One Day Maritime History Conference, 1 June 1996.
Brooks, J. 'The Battlecruiser Duel: British and German Gunnery at Jutland'.
Gordon, G.A.H. 'The Battle of Jutland'.
Grove, E. 'Who Really Did Win, Then?'.
Sumida, J.T. 'The Neptune Factor: Mahan Reassessed'.
Sumida, J.T. and Lambert, N. 'Jutland Reconsidered'.

Society for Nautical Research/Kings College, London

Gordon, G.A.H. 'Historians, Navigation and the Battle of Jutland', 13 March 1997.

Hill, J.R. 'Royal Navy Policy since 1945', 1 February 1996.
Till, G. 'The British Approach to Amphibious Operations', 30 January 1997.

National Maritime Museum (NMM), Greenwich/University of Westminster

'Adapting to Change: The Royal Navy and the Maritime Industries, 1815–1990', One Day Maritime History Conference, 31 October 1998.
Harding, R., 'The Royal Navy and Amphibious Operations, 1919–1939'.

University of Lancaster/Royal Navy/BAE

Royal Navy Submarine Centenary Conference, 27–29 September 2000.
Gardner, W.J.R. 'Two Committees, Three Submarine Classes and 31 Hulls: The "R", "K" and "M" class Submarines'.
Grove, E. 'British Submarine Policy in the Inter-War Period 1918–1939'.

ARTICLES AND CHAPTERS

Allen, M. 'The Foreign Intelligence Committee and the Origins of the Naval Intelligence Department of the Admiralty', *The Mariners Mirror*, Vol. 81, No. 1, February 1995, pp. 65–78.
Baugh, D. 'Admiral Sir Herbert Richmond and the Objects of Sea Power', in J. Goldrick and J. B. Hattendorf (eds), *Mahan is Not Enough – Conference on the Work of Sir Julian Corbett and Sir Herbert Richmond* (Naval War College Press, Newport, RI, 1994), pp. 13–49.
Bell, C. 'Our Most Exposed Outpost: Hong Kong and British Far Eastern Strategy 1921–1941', *Journal of Military History*, Vol. 60, No. 1, January 1996, pp. 61–88.
—— 'The Royal Navy, War Planning and Intelligence Between the Wars', in J. Siegel and P. Jackson (eds), *Intelligence and the International System, 1870–1970* (Praeger, New York, forthcoming).
—— 'How Are We Going to Make War? Admiral Sir Herbert Richmond and British Far Eastern War Plans', *Journal of Strategic Studies*, Vol. 20, No. 3, September 1997, pp. 123–41.

Best, A. 'Constructing an Image: British Intelligence and Whitehall's Perception of Japan, 1931–1939', *Intelligence and National Security*, Vol. 11, No. 3, July 1996, pp. 403–23.

Bitner, J.F. 'The Royal Marines and Amphibious Warfare in the Inter War Years', *Journal of Military History*, Vol. 55, 1991.

Breakfield, Major W., USMC. Letter in reply to Pierce, *Proceedings*, Vol. 121, April 1995, *US Naval Institute Proceedings*, Vol. 121, October 1995, p. 27.

Bromley, A.G. 'British Mechanical Gunnery Computers of World War II', Technical Report 223, Basser Department of Computer Science, University of Sydney, January 1984.

Brown, D.K. 'Gunnery at Jutland', *The Mariners Mirror*, Vol. 73, No. 2, May 1987, p. 209.

Bullen, J. 'The Royal Navy and Air Power: the Projected Torpedo-Bomber Attack on the High Seas Fleet at Wilhelmshaven in 1916', *Imperial War Museum Review*, No. 2, 1987, pp. 71–7.

Callo, R.A.J.F., USN(R). 'Finding Doctrine's Future in the Past', *US Naval Institute Proceedings*, Vol. 122, October 1996, pp. 64–6.

Cooper, Lieutenant Commander G., USN. Letter in Reply to Pierce, *Proceedings*, Vol. 121, April 1995, *US Naval Institute Proceedings*, Vol. 121, August 1995, p. 21.

Cowman, I. 'An Admiralty Myth – The Search for a Far Eastern Base before the Second World War', *Journal of Strategic Studies*, Vol. 8, No. 3, September 1987.

—— 'Defence of the Malay Barrier? The Place of the Philippines in Admiralty Naval War Planning 1925–1941', *War in History*, Vol. 3, No. 4, 1996, pp. 398–417.

Geddes, D. 'The Mandate for Yap', *History Today*, Vol. 43, December 1993, pp. 32–7.

Gooch, J. 'Hidden in the Rock: American Military Perceptions of Great Britain, 1919–1940', in L. Freedman, P. Hayes and R. O'Neill (eds), *War, Strategy and International Politics* (Clarendon, Oxford, 1992), pp. 155–73.

Gordon, G.A.H. 'The British Navy 1918–1945', paper from the Symposium, *Navies and Global Defence*, Royal Military College of Canada (March 1994).

—— 'The Doctrine Pendulum', in *The Hudson Papers*, Vol. I: *Lectures to the University of Oxford* (Ministry of Defence (Navy), London, 2001).

Gow, I. 'Anglo-Japanese Naval Relations prior to 1931', London School of Economics, International Studies, 1985.

Gueritz, Admiral E.F. 'Gunnery at Jutland', response to D.K. Brown, *The Mariners Mirror*, Vol. 73, No. 2, May 1987, p. 209.

Hirama, Rear Admiral Y. 'Interception–Attrition Strategy from the Washington Treaty to the Battle of the Philippine Sea', in *Les marines de guerre du dreadnought au nucleaire. Actes du Colloque international Paris, ex-Ecole Polytechnique*, 23–25 November 1988, pp. 391–412.

Hoffman, Lieutenant Colonel F.G., USMCR. 'A Littoral Leader', *US Naval Institute Proceedings*, Vol. 122, October 1996, pp. 60–6.

Hone, T.C. Letter in reply to Roncolato, *Proceedings*, Vol. 122, June 1996, *US Naval Institute Proceedings*, Vol. 122, October 1996, p. 19.

Hughes, W. 'Manoeuvring Past Manoeuvre Warfare', *US Naval Institute Proceedings*, Vol. 122, March 1996, p. 16.

Ikeda, K. 'Anglo-Japanese Naval Relations', *The Mariners Mirror*, Vol. 71, No. 4, November 1985, pp. 444–5.

Kuramatsu, T. 'The Geneva Naval Conference of 1927: The British Preparation for the Conference, December 1926 to June 1927', *Journal of Strategic Studies*, Vol. 19, No. 1, March 1996.

Lambert, N. 'Admiral Sir John Fisher and the Concept of Flotilla Defence, 1904–1909', *Journal of Military History*, Vol. 59, October 1995, pp. 639–60.

Lind, W. Letter in reply to Pierce, *Proceedings*, Vol. 121, April 1995, *US Naval Institute Proceedings*, Vol. 121, September 1995, p. 23.

Liske, Lieut H.P., USN. Letter in reply to Pierce, *Proceedings*, Vol. 121, April 1995, *US Naval Institute Proceedings*, Vol. 121, July 1995, p. 19.

Lyle, C. 'Jutland or a Second "Glorious First of June"?', *The Mariners Mirror*, Vol. 82, No. 2, May 1996, pp. 190–9.

McBride, K.D. '1915 – Task Forces that Never Were', *The Mariners Mirror*, Vol. 79, No. 1, February 1993, p. 90.

McDonald, K.J. 'The Rainbow Plans and the War against Japan', in *Symposium on The Second World War in the Pacific: Plans and Reality*, National Maritime Museum Monograph No. 9, 1974, pp. 20–9.

MacGregor, D. 'The Use, Misuse and Non-Use of History: The Royal Navy and the Operational Lessons of the First World War', *Journal of Military History*, Vol. 56, No. 4, 1992.

Mahnken, T.G. '"Gazing at the Sun": The Office of Naval Intelligence and Japanese Naval Innovation 1918–1941', *Intelligence and National Security*, Vol. 11, No. 3, July 1996, pp. 424–41.

Murfett, M. 'Living in the Past: A Critical Re-examination of the Singapore Naval Strategy 1918–1941', *War in Society*, Vol. 11, No. 1, May 1993, pp. 77–97.

Pehl, Captain C.E., USN(R). 'Through a Glass, Darkly', *US Naval Institute Proceedings*, Vol. 122, September 1996, pp. 54–7.

Pierce, Commander T.C., USN. 'Taking Manoeuvre Warfare to Sea', *US Naval Institute Proceedings*, Vol. 121, April 1995, pp. 74–7.

Primrose, B.N. 'Australian Naval Policy 1919–1942: A Case Study in Empire Relationships', Thesis submitted for the Degree of Doctor of Philosophy in the Department of International Relations, Australian National University, September 1974.

Ranson, E. 'British Defence Policy and Appeasement Between the Wars 1919–1939', The Historical Association, 1993 (Pamphlet).

Reynolds, C. 'The US Fleet-in-Being Strategy of 1942', *Journal of Military History*, Vol. 58, January 1994, pp. 103–18.

Rodger, N.A.M. 'Training or Education: A Naval Dilemma Over Three Centuries', *The Hudson Papers, Vol. I: Lectures to the University of Oxford* (Ministry of Defence (Navy), 2001).

Roncolato, Commander G.D., USN. 'Methodical Battle', *US Naval Institute Proceedings*, Vol. 122, February 1996, pp. 32–3.

—— 'Taking Manoeuvre Warfare to Sea', *US Naval Institute Proceedings*, Vol. 122, March 1996, p. 20.

—— 'No Time to Rest', *US Naval Institute Proceedings*, Vol. 122, June 1996, pp. 30–4.

Roskill, Captain S.W. 'The Second World War in the Pacific – Plans and Reality: The British Point of View', paper delivered at Symposium, *The Second World War in the Pacific: Plans and Reality*, Monograph No. 9, NMM, 1972, pp. 9–19.

—— 'Naval Policy Between The Wars', National Maritime Museum Monograph, 1977.

Saxon, T. 'Anglo-Japanese Naval Co-operation 1914–1918', *Naval War College Review*, Winter 2000 (United States Naval War College, Annapolis, MD).

Schleihauf, W. 'A Concentrated Effort: Royal Navy Gunnery Exercises at the End of the Great War', *Warship International*, No. 2, 1998, pp. 117–39.

Shiflett, C.R. 'The Royal Navy and the Question of Imperial Defence East of Suez, 1902–1914', *Warship International*, No. 4, 1995, pp. 353–66.

Storry, R. 'The Greater East Asia War as the Japanese Saw It', paper delivered at Symposium, *The Second World War in the Pacific – Plans and Reality*, National Maritime Museum, 1972, pp. 3–8.

Sumida, J.T. 'British Capital Ship Design and Fire Control in the *Dreadnought* Era: Sir John Fisher, Arthur Hungerford Pollen and the Battlecruiser', *Journal of Modern History*, June 1979.

—— '"The Best Laid Plans": The Development of British Battle Fleet Tactics 1919–1942', *International History Review*, Vol. 14, November 1992, pp. 661–880.

—— 'British Naval Operational Logistics 1914–1918', *Journal of Military History*, Vol. 57, No. 3, 1993.

—— 'Sir John Fisher and the *Dreadnought*: The Sources of Naval Mythology', *Journal of Military History*, Vol. 59, October 1995, pp. 619–37.

—— 'Fisher's Naval Revolution', *Naval History*, Vol. 10, August 1996, pp. 20–6.

Tsokhas, K. 'Anglo–Australian Relations and the Origins of the Pacific War', *History*, Vol. 80, October 1995, pp. 400–20.

Whitby, M. 'In Defence of Home Waters: Doctrine and Training in the Canadian Navy during the 1930s', *The Mariners Mirror*, Vol. 77, No. 2, May 1991, pp. 167–77.

Wright, A. 'Australian Carrier Decisions: The Acquisition of HMA Ships *Albatross*, *Sydney* and *Melbourne*', *Papers in Australian Maritime Affairs*, No. 4, Canberra 1998.

Index

ABC-1 Staff Agreement, 216–17
Abo, Japanese Navy Minister, 192
Abyssinian Crisis, 41, 44, 101, 213
Aden, 57
Admiralty Fire Control Table, 14, 128–9
Admiralty, the, 30; and the Anglo-Japanese Alliance, 25; and Australia, 7; and the battleship building holiday, 27; Cabinet power over, 10; cruiser requirements, 27–8, 29; and the Eastern Fleet, 219–20, 220–1; and a Far Eastern Fleet, 8–9; Far Eastern risk assessment, 49; and Hong Kong, 102; and the Japanese threat, 25; and naval aviation, 12; naval policy, 11; and naval spending, 21, 44; and naval strength, 13; and the need for a naval building programme, 25; and the need for an imperial navy, 22–3; Pacific strategic focus, 23; Plans Division, 48–9, 54, 56, 68, 92, 93, 97, 159, 231; rejects Jellicoe's report, 23; strategy, 23–4, 230–1, 232; and the Ten Year Rule, 26, 27, 29–30, 233; and US assistance, 117; war plans, 14, 30–1, 45–6, 48–9, 49–50, 56, 63–4, 97–8, 101–2, 159, 217, 230; and the Washington Naval Treaty, 7, 43
advanced base, 57, 58
Adventure, HMS, 14, 91
Agincourt, HMS, 59, 164
air power, 12, 69, 79, 81, 150, 152–3, 170, 238–42; Imperial Japanese Navy and, 191–2, 203–5, 205–8, 241
air reconnaissance, 166, 171

aircraft, 3, 12, 45, 50, 115, 131, 153, 172–3, 216, 239, 241; bombs, 150, 152, 173; Imperial Japanese Navy, 192–3, 209; tactical role, 132, 136–7, 151, 171, 173, 205, 209; torpedo bombers, 149, 150–1, 151–2, 165, 173
aircraft carriers, 12, 29, 45, 115, 119–20, 121, 137, 150, 172–4, 206, 241–2; Imperial Japanese Navy, 79, 166, 189, 192, 193, 195, 204–5, 206–7, 208–9; tactical role, 136, 151–2, 204–5, 208–9, 239; US Navy, 206; vulnerability of, 166
Alboran Island exercise, 167–70
Alexandria, 101, 110
allied naval forces, 121
Althan, Captain E.H., 194
Anderson, Sir Alan, 13
Anglo-German Naval Treaty, 40–1, 44, 100–1, 213, 243
Anglo-Japanese Alliance, 3, 5, 19, 21, 22, 25, 49
anti-submarine warfare, 3, 14–15, 245
appeasement, 40–1
Argus, HMS, 151, 172, 206, 222
Ark Royal, HMS, 222
arms limitation, 32, 38, 41–2, 42–3
ASDIC, 14
Ashmore, Admiral Sir Edward, 163
Assistance, HMS, 67
Atlantic Conference, 118, 217
Atlantic Fleet Battle Instructions, 130, 135–43
Atlantic Ocean, 218
Atlee, Clement, 221

INDEX

Australia, 64, 68, 94; and the Anglo-Japanese Alliance, 5; defence of, 109–10, 122n19; defence policy, 7; and the Flag Officers meeting, 72n27; reassurance of, 104, 109–10; and the US Navy, 7

Backhouse, Admiral Sir Roger, 9, 104, 107, 109–10, 112, 194, 237, 243–4
Bacon, Admiral Sir Roger, 175–6
Baldwin, Stanley, 21, 29, 39
Barham, HMS, 120, 129, 222
Bartolomé, Rear Admiral de, 49
battle cruisers, 137, 140, 154
battle fleet strengths, 20 (map)
Battle Instructions, 3, 12, 45, 130–1, 179, 236, 237–8, 247
battleships, 3, 12, 42, 149, 149–50, 233; bomb hits on, 152; building holiday, 26; costs, 72n32; IJN, 120–1, 195, 196; oil consumption, 93; Royal Navy, 24, 103–4, 120–1, 194, 242; tactical role, 131, 139–40
Beatty, Admiral Sir David, 2, 30, 49, 98, 236; command style, 124, 126, 132, 236; and the Japanese threat, 25; and the Naval Estimates, 10, 25; and naval threats, 21–2; on the need for a naval building programme, 24; and a Peace Fleet, 60–1; and Singapore, 51, 62–3; strategic doctrine, 54
Bee, HMS, 103
Birkenhead Committee, 29
Bonar Law Enquiry, 49, 233
British Naval Staff History of the Second World War in the Pacific, 184
Brock, Rear Admiral Osmond de B, 49, 51, 53
Brownrigg, Admiral Sir Studholme, 176

Cabinet, the, 14; abandons the Ten Year Rule, 100; and the Japanese threat, 11, 30; power over the Admiralty, 10
Chamberlain, Austen, 234
Chamberlain, Neville, 100, 104
Chatfield, Admiral Lord, 30, 40, 44, 98, 99, 100, 101, 179–80, 213, 238; advocates the Anglo-German Naval Agreement, 100–1; challenges small squadron idea, 111–12; on the Mediterranean, 112
Chiefs of Staff, 103, 114
Chiefs of Staff Committee, 40, 100
China, 5, 87, 214
China Incident, the, 193
Churchill, Winston, 5, 21, 116, 118–19, 217; as Chancellor of the Exchequer, 29; and the Eastern Fleet, 217–19, 220–1; perception of the Japanese threat, 10–11, 17n10, 29, 30, 218, 221–2, 234
coal, 91, 93
Coastal Command, RAF, 152
collective security, 26, 41
Colombo, Ceylon, 53, 57
Commander-in-Chief, East Indies Station, 58
Committee of Imperial Defence, 53–4, 104–6, 243
communications, 3, 124, 126, 171–2
convoys, 166, 167
Coral Sea, Battle of the, 209
Cornwall, HMS, 1, 215
Courageous, HMS, 136, 151, 170, 172
Cowan, Rear Admiral, 130
Cricket, HMS, 103
cruisers, 8, 27–9, 39, 42, 43, 70, 98, 149, 242; Imperial Japanese Navy, 28, 31, 32, 38, 189, 191–2, 196, 196–7; tactical role, 131, 137, 140, 141, 154, 197; US Navy, 31, 32, 38, 189–90, 191
cruising formations, 136 (table)
Cunningham, Admiral Andrew Browne, 104–5, 112, 175
Czechoslovakia, invasion of, 215

Danae, HMS, 215
Danckwerts, Captain V.H., 112–13
Darwin, 1, 225
Dauntless, HMS, 215
Decoy, HMS, 214
Defence Requirements Committee, 40, 42, 99
'Despatch of a Fleet to the Far East, The' (memo), 105
destroyers, 3, 120, 166–7, 197, 242–3; tactical role, 131–2, 140, 141–2
Dewar, Captain, 22
dockyard facilities, 91–2
Dominions, the, 6, 7, 56; defence responsibility, 67–8; and an imperial navy, 23; reassurance of, 104
Domvile, Captain, 22, 49
Dorsetshire, HMS, 1, 215

INDEX

Drax, Admiral R.A.R. Plunkett-Ernle-Erle, 243–4, 247; on leadership, 153–4; on manuals, 178–9; parts of a battle, 143; tactical principles, 130–3, 180–1, 236, 237; war plan, 107–10
Dreyer, Admiral Frederick, 106–7, 130
Dreyer Table, the, 127–8
Duke of York, HMS, 218, 219
Durban, HMS, 215, 226
Dutch Navy, 223 (map)

Eagle, HMS, 151, 170, 172, 206, 214, 215, 222
Eden, Anthony, 221
Egerton, Rear Admiral, 89–90
Eliot, Major Fielding, 224
Ellington, Air Chief Marshal, 40
Emerald, HMS, 216
Enterprise, HMS, 216
Europe, effect of war in, 115–16
exercises, 3, 144–9, 151, 158–9, 166, 238, 246; Alboran Island exercise, 167–70; Ashmore on, 163; combined air and surface, 166; conclusions from, 177–81; Exercise EA, 164; Exercise MI, 172; Exercise MQ, 166–7; Exercise MU, 164–6; Exercise MU2, 167–70; Exercise MZ, 163, 170–2; Exercise NASF, 164; Exercise PZ, 180–1; Fisher's exercise, 174–6; Imperial Japanese Navy, 205–8; Salsette Island, 51, 78–81, 90, 235; Scheme J1, 160–3; Scheme X2, 159–60
Eyres-Monsell, Sir Bolton, 240–1

Field, Admiral Sir Frederick, 98, 98–9
fire control, 124, 127–9, 138–9
Fisher, Admiral W.W., 174–6; as Vice-Admiral, 30
Fisher, Sir Warren, 40
Flag Officers Conference, Penang, 67, 68, 68–71, 72n27
Fleet Air Arm, 115, 152, 239, 240–1
fleet strength, 82–3
Foreign Office, the, 5, 232
Formidable, HMS, 218, 222
France, fall of, 215–16, 245
Franco-Italian antagonism, 43
French Indo-China, 216, 244, 245
French Navy, 104, 111, 215
fuel oil, 53, 57, 59, 66–7, 92–3; requirements, 51, 55 (table), 56, 91; reserve programme, 43; storage proposals, 61 (table)
Furious, HMS, 25, 136, 151, 172, 222
Furutaka class, 28, 196

Genda, Minoru, 205
Geneva Conference (1927), 31–2, 189–90
geographical position, importance of, 94
German Navy, 41, 99
Germany, 41, 99, 243; threat of, 40, 98, 99, 100, 101
Gibraltar, 60
Glorious, HMS, 136, 151, 172
Gordon, Andrew, 2, 236
Grand Fleet Battle Instructions, 126
Grand Fleet Battle Orders, 131
Grand Fleet Manoeuvring Orders, 126
Grant, Admiral Sir William Lowther, 4, 22
Great Britain: challenge of protecting the Empire, 7–8; defence responsibility, 67–8; defence spending, 100–1; effect of war with Japan, 76; Far Eastern interests, 97–8; foreign policy, 40; naval building programmes, 8, 24, 29–30, 32, 42, 43, 45, 233; and naval parity with the US, 4; naval spending, 43; oil reserves, 53; reaction to the Washington Treaty, 7
Great War (1914–18), 2–3, 19, 21
gunnery, 3, 81, 126, 129–30, 138–9; anti-aircraft, 148–9, 239–40; comparison of strength, 89, 95n31; damage assessment, 145, 146 (table); danger zones, 177; effects, 146 (table); firing systems, 126–7; hit rates, 124, 144–5, 145 (table); hits per minute, 145, 147 (table), 148; at Jutland, 124

Hainan, 104
Hankey, Sir Maurice, 40
Hankow, 99
Hart, Admiral Thomas, 225
Hawkins, HMS, 58
Henderson, Admiral Sir Reginald, 173–4, 240
Hermes, HMS, 1, 121, 151, 172, 206, 222
Hong Kong, 22, 26, 57, 70, 75, 79, 80, 84–5, 88, 102, 216; defences, 61–2; and exercises, 160, 161; fall of, 1; importance of, 106

INDEX

Hood, HMS, 24, 38, 120, 129, 131, 149, 242
Hughes, William, 5–6

Illustrious, HMS, 218, 222
Illustrious class, 115
Imperial Conferences (1909, 1921), 5
Imperial Japanese Navy, 1, 6, 7, 58; and air power, 191–2, 203–5, 205–8, 241; aircraft, 192–3, 209; aircraft carriers, 79, 166, 189, 192, 193, 195, 204–5, 206–7, 208–9; battleships, 120–1, 195, 196; comparison with Royal Navy, 120–1; cruisers, 28, 31, 32, 38, 189, 191–2, 196, 196–7; destroyers, 197; effectiveness, 120; exercises, 205–8; heavy cruisers, 188, 189; links with the Royal Navy, 19, 183; mobilization, 224, 224–5; naval air arm, 12; strategy, 184–6; strength, 24, 43, 101, 117, 183, 186, 194–5, 195 (table), 223 (map); submarines, 189, 196, 197; tactics, 195–6, 203, 208–9; and the Washington Naval Treaty, 26
India, 75, 86; and the Anglo-Japanese Alliance, 5
Indian Ocean, 1, 227–8
Indomitable, HMS, 121, 222
initiative, 126, 132–3, 135, 179, 180, 237; use of, 139, 140, 154
intelligence, lack of, 99–100
Invergordon Mutiny, the, 13–14
Iron Duke, HMS, 38, 93
Iron Duke class, 38, 95n31, 131
Italy, 108; invasion of Abyssinia, 41, 101; naval forces, 112; threat of, 40, 44, 98, 99, 101–2, 110

Japan, 1; the 8:8 Plan, 19, 186–7, 187–8, 188; aims, 184–5; difficulty to blockade, 75, 76–7; dockyards, 24–5; and the Great War, 19; incidents with, 214–15; national concerns, 184; naval bases, 56, 79, 83; naval building programmes, 3, 5, 21, 28, 189, 190–1; naval replenishment programmes, 192–4; scheme to attack Singapore, 100; strategic preparations, 183; strategy, 188–9, 209–11; threat of, 4–5, 6, 7, 8, 10–11, 17n10, 22, 23, 25, 30, 40, 63–4, 98, 105–6, 183, 231, 234–5, 246; trade, 56, 75; and the Washington Naval Treaty, 186–8
Japan Must Fight Britain (Tota), 185
Java Sea, Battle of the, 1, 210, 227
Jellicoe, Admiral Sir John, 2, 39; command style, 124, 134, 236; Dominion Mission, 6, 22–3; and the Japanese threat, 23, 183, 231; success at Jutland, 126
Johore Straits, 50
Joint Planning Committee, 114–15
Joint Planning Staff, 217
Jutland, Battle of, 2, 16n3, 45, 124, 126, 130, 203, 236, 237

Kanji, Kato, Admiral, 188
Keelung, 214
Kent class, 28
Keyes, Admiral Roger, 130, 165
King George V, HMS, 42
King George V class, 27, 44, 45, 47n30, 129, 217, 242
Kirk, Captain, 116

Ladybird, HMS, 103
Lambert, G., 246–7
Larken, Commander, 19, 21
League of Nations, 10, 26, 32, 41
Leander class, 38
Lexington, USS, 206
Libya, 110
lines of communication, 57, 58, 66, 72n46
Lion class, 44
Liverpool, HMS, 215
Lloyd George, David, 4, 5
logistics, 51, 57, 63, 76, 81–2
London Naval Conference (1930), 38–9, 98, 191–2, 208; (1936), 41–2
Long, Lord, 22

MacDonald, Ramsay, 21, 28, 32, 39, 98
Madden, Admiral Sir Charles, 30, 38, 98, 135–6, 236
Malacca, Straits of, 69–70, 164–5
Malaya, HMS, 129, 222
Malaya, Japanese invasion of, 226, 244
Malaya Barrier, the, 64–6
Malta, 60, 101, 110
Manchuria, Japanese invasion of, 41, 192
Marder, Arthur, 15
Marriott, W.F., 24–5

270

INDEX

Matapan, Battle of, 218
May, Sir George, 13
Mediterranean Sea, the, 8, 59, 60, 101, 101–2, 104, 107, 112, 214, 218; Pound's strategy, 110–11, 113, 119
Menzies, Robert, 119
Midway, Battle of, 209
Millar, E.W.H., 54
Milne, Lord, 99
Mobile Naval Base Defence Organization, 59, 91, 160, 161, 164
Montgomery-Massingberd, Field Marshal, 40
morale, 143
Munich Crisis, 243

Nagato class, 95n31
Nancowry Harbour, 81
naval balance (1941), 223 (map)
naval bases, 22, 97
Naval Estimates, 5, 7, 10, 21, 22, 25, 28–9, 39
Naval Staff, 15
Naval Staff College, 130
Naval War Manual, 178–9, 238
naval warfare, 9
Nelson, HMS, 14, 27, 29, 38, 115, 129, 176, 217, 242
Nelson class, 95n31, 131
Netherlands East Indies, 53, 70
New Zealand, 5, 64, 68, 94
night actions, 3, 142–3, 166, 174–7, 179–80, 203; Japanese proficiency in, 196
Noble, Vice-Admiral Sir Percy, 109
Norfolk, HMS, 14
Northumberland, HMS, 38
Norway campaign, 45

oilers, 93
One Power Standard, the, 11, 27, 51, 67, 68

Pacific Ocean, 4, 5–6; allied naval power in, 227–8; Japanese Great War naval deployment in, 19; Japanese trade, 56; strategic importance, 19, 21
Panay, USS, 103, 114
Panzerschiff (pocket battleships), 99
Paracel Islands, 183
passage time, 51, 53, 90–1
Peace Fleet, 60–1, 98, 244
Pearl Harbor, 60, 118, 203, 209, 213, 226

Pegasus, HMS, 136
Penang, Flag Officers Conference, 67, 68, 68–71, 72n27
Pescadores Islands, 84
Philippines, the, 97–8
Phillips, Rear Admiral Tom, 119, 225, 226, 227
Plan X, 106
Plan Y, 106–7
Port Darwin, 70
Port Said, 57
Post War Questions Committee, 233
Pound, Captain Dudley, 28, 68, 101, 118, 222; on the Anglo-German Naval Treaty, 101; and the Eastern Fleet, 219–20, 220; Mediterranean strategy, 110–11, 113, 119; and US assistance, 116
preparation strategy, 158
Prince of Wales, HMS, 1, 42, 111, 119, 121, 203, 210, 222, 224, 226–7

Queen Elizabeth, HMS, 120, 129, 222, 242
Queen Elizabeth class, 27, 38, 89, 115, 131, 149, 177, 217

R class, 38, 89, 129, 131, 149, 217, 227–8
Ramillies, HMS, 118, 120, 129, 217
Range Clocks, 126–7, 129
rangefinders, 124, 127
Rangoon, 57
Red Sea, 53
refuelling facilities, 53
Renown, HMS, 38, 115, 129, 217, 218, 242
Repulse, HMS, 1, 38, 111, 120, 121, 129, 131, 149, 203, 210, 218, 222, 224, 225, 226–7, 242
Repulse class, 27
Resolution, HMS, 120, 129, 217
Revenge, HMS, 120, 129, 217
Richmond, Admiral Sir Herbert, 9, 9–10, 50, 68, 130, 235; attack on Singapore scenario, 74–5; and British gun superiority, 89; criticism of War Memorandum (Eastern), 71, 74–94, 97, 247; on economic blockade, 76–7; flaws in War Memorandum (Eastern), 77–8; on a fleet action, 74, 75, 77; guiding principle, 133; and Hong Kong, 80, 84–5; on India, 75, 86; naval

271

INDEX

objectives, 87; proposals, 86–9; and Slade, 85–6; tactical principles, 133–43, 236–7
River Plate, Battle of the, 244–5
Rodney, HMS, 14, 27, 29, 38, 115, 129, 176, 217, 242
Roosevelt, Franklin D., 118, 217
Ross, Captain, 99
Royal Air Force, 12, 152–3
Royal Australian Navy, 38, 39, 49, 68
Royal Naval Air Service, 152
Royal Naval College, 159
Royal Navy, 15, 21; aircraft carriers, 206, 241–2; Atlantic Fleet, 25, 60, 180; and aviation, 12; battleships, 24, 103–4, 120–1, 194, 242; belief in superiority, 123; China Fleet, 25, 59; China Squadron, 49, 62, 84, 85, 103, 120, 211n11, 214, 215, 216, 245; comparison with Imperial Japanese Navy, 120–1; cruiser squadrons, 9; cruisers, 31, 32, 39, 98, 189, 191, 242; culture, 15; debate on future development, 3; decline in strength, 39–40; decline of vessels, 32; destroyers, 242–3; deterrent, 118; disposition, 59–61; East Indies Fleet, 1; East Indies Squadron, 85; Eastern Fleet, 217–21, 222, 227; Far Eastern Fleet, 59; force at Singapore, 120; Force H, 219, 222; Force Z, 1, 121, 225–7; and the Great War, 2; Home Fleet, 25, 219; home waters defence needs, 105; inability to operate in two hemispheres, 6; inter-war improvements, 247; lessons of the Great War, 2–3; links with Imperial Japanese Navy, 19, 183; Mediterranean Fleet, 25, 39, 41, 90–1, 110–11, 179–80, 180; modernization, 27, 44–5, 101; naval aviation doctrine, 239; naval pay cuts, 13–14; need for Far Eastern naval base, 49; officers, 15–16; organization, 9–10; Pacific Fleet, 91; potential enemies debate, 3–5; proposed battle fleet, 103–4; Reserve Fleet, 25; self-confidence, 11, 45; self-image, 132; situation (1941), 222; Special Service Squadron, 68; strategic priorities, 110; strength, 4, 6–7, 8, 21–2, 27, 38, 39, 41, 43, 58, 105, 115, 215–16, 218–19, 223 (map); 234–5; superiority to German fleet, 41; superiority to Imperial Japanese Navy, 23–5; war plan, 12–13; and the Washington Naval Treaty, 7, 8
Royal New Zealand Naval Division, 38
Royal Oak, HMS, 129
Royal Sovereign, HMS, 120, 129, 217
Rules of the Game, The (Gordon), 2

Salsette Island, invasion exercise, 51, 78–81, 90, 235
Saratoga, USS, 206
Scarab, HMS, 103
secret bases, 66–7
Shanghai crisis, 214
Singapore, 6, 9, 12, 17n12, 17n13, 43, 53–4, 232, 246; to be developed as a naval base, 23; defences, 50–1, 54, 100, 107, 117, 165–6, 214, 216; distances from, 83; and exercises, 160, 161; force size, 111–13, 120; fuel oil reserves, 55 (table), 67, 93; garrison, 51; importance of, 29, 59, 61, 62–3, 64, 97, 106, 247; Japanese landing threat, 50–1; justification for base, 94; naval force, 224; oil storage capacity, 53, 61 (table); Period Before Relief, 104, 216; relief exercise, 167–70; relief of, 177; Richmond on, 74–5; role, 57; security, 44; surrender of, 1; threat from if captured, 87–8; and the Washington Naval Treaty, 26
Singapore Defence Conference (1940), 216
Slade, Admiral Edmund, 83, 84, 85–6
Somerville, Admiral James Fowler, 1, 227–8
Spratley Islands, 104
Staff College Exercises, 3
Stewart, Lieutenant Commander R.R., 62
strategic centre of gravity, 19, 21
strategic objectives, 9
strategic situation, 113–14, 119–20, 213–17
Strategical Appreciation Sub-Committee, 104–5, 111
strategy, 23–4, 32, 48, 58, 63–6, 76, 92, 93–4, 123, 230–1, 231–2, 234, 245–9; danger of generalizations, 75; effect of Washington Naval Treaty, 59;

INDEX

Japanese, 184–6, 188–9, 209–11; shortcomings, 41
submarines, 3, 42, 69, 189, 196, 197; development, 142; tactical role, 132, 140, 142
Suez Canal, 44, 90, 103, 111
Sunda Strait, 69
supply arrangements, 55 (table)
Surrey, HMS, 38

tactical doctrine, 83
Tactical Manual (1922), 133, 135
Tactical Manual (1925), 123
Tactical School, 130
tactics, 3, 12, 123–4, 133–4, 153–5, 180, 235–8, 244–5; Drax's tactical principles, 130–3, 180–1, 236, 237; gunnery, 129–30; Imperial Japanese Navy, 195–6, 203, 208–9; Richmond's tactical principles, 133–43, 236–7
Takahashi, Rear Admiral Korekizo, 184, 206
Taniguchi, Admiral, 208
Taranto, 238
Ten Year Rule, 26, 27, 29–30, 100, 159, 233
theatre of operations, size of, 188–9
Tientsin, 113, 215, 244
Tiger, HMS, 38, 82, 89, 93
Tiger class, 95n31
torpedoes, effect of, 149
Tota, Lieutenant Commander Ishimaru, 185
trade warfare, 32, 94, 159–60
training, 124, 130, 132–3, 232
Treasury, the, 21, 22, 27, 42
Trenchard, Air Chief Marshal Hugh, 50
Trincomalee, 91, 220
Tufnell, Captain, 194
two-power standard, 40–1

UK/US Staff Conference, 117–18
United States Navy, 1, 4, 6, 7, 114, 116–17; aircraft carriers, 206; Asiatic Fleet, 211n11, 223 (map); cruisers, 31, 32, 38, 189–90, 191; naval air arm, 12; Pacific Fleet, 117, 118, 121, 217, 223 (map), 244; strength, 43, 58
United States of America, 97; and the defence of Singapore, 117; and the Eastern Fleet, 220; involvement, 116–17; and the Japanese threat, 117–18; naval building programmes, 3, 4, 21; naval spending, 43; as potential enemy, 3–4, 6; statement of intentions, 216–17

Valiant, HMS, 14, 120, 129, 222, 242
Vansittart, Sir Robert, 40
Vernon, HMS, 124
Victorious, HMS, 222
Victory, HMS, 124

War Cabinet, meeting of 17 October 1941, 221–2
War College, 130
War Memorandum (Eastern), 2, 3, 8, 21, 65 (map), 117, 230–1, 244; 1927 revision, 91–2; 1931 revision, 92–3, 97; 1937 revision, 102–3; and the Flag Officers Conference, 68–71; importance of, 248; naval forces, 63–4; objective, 66, 123–4; Phase I, 89, 90; Phase II, 89; Phase III, 83, 89, 89–90; Phases, 64–6, 235; as 'preparation strategy', 48–9; Richmond's criticism of, 71, 74–94, 97, 247; Richmond's flaws in, 77–8; Richmond's proposals, 86–9; route to the East, 52 (map); and Singapore, 64; strategic objectives, 9; value of, 246
War Plan Orange, 114, 118
Warspite, HMS, 120, 129, 225, 227, 242
Washington Conference and its Effects upon Empire Naval Policy and Co-Operation, The (memorandum), 68
Washington Four Power Treaty, 64
Washington Naval Treaty (1921), 1, 6–7, 8, 26, 27, 43, 49, 58, 94, 186–8, 233; limit on fortifications, 61–2